UTAH TRILOBITES
ROBERT HARRIS COLLECTION

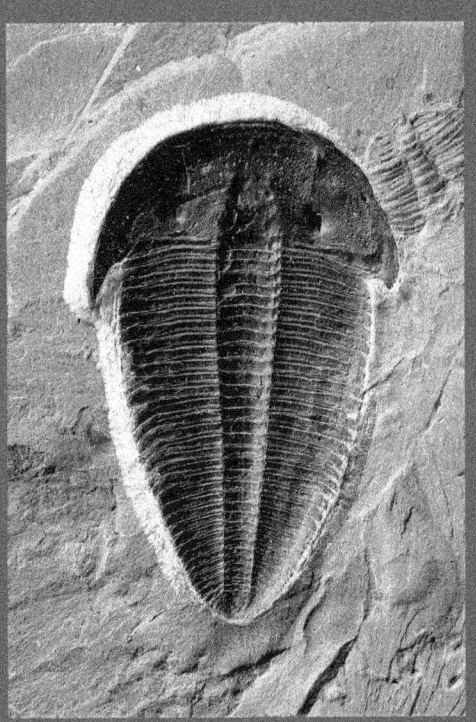

BY MARK HARRIS

UTAH TRILOBITES

Cover Photograph: The trilobite *Altiocculus harrisi* is one of many species of fossil trilobites found in the Cambrian-age sedimentary formations of the House Range in Western Utah. This impressive specimen is a well-preserved fossil trilobite in the collection of Robert L. Harris of Delta, Utah. Since Robert discovered one of the first nearly perfect specimens of this trilobite, it has been given the name *"harrisi"*.

Utah Trilobites: is self-published by Mark F Harris

©2019 by Mark F Harris
All Rights Reserved
Printed in the United States of America

For additional copies of this book
mlharris@pacbell.net, or
Internet: Utah Trilobites by Mark Harris

ISBN: 978-1-716-51155-4

Also by Mark F Harris:
The Harris Solution to Rubik's Cube
A Distant Place
The Missionary Journals of Edward Daniel Harris
Letters from China
Mark's Little Joke Book
Letters from Serbia (With Luree Condie Harris)
Reflections: Selected Writings of Mark F. Harris
Tay Attent ion
Photographic Reflections
Photographic Patterns
Photographic Composition
Southern Utah, Land of Color
Poetry and Potpourri (Editing)
My First Eighty Years
Ted Harris: Area Giant
Life Sketch of Henry Forster
Harris Family Ancestry
John Umpstead Rencher Story

Cover design and text formatting: Mark F Harris
Trilobite Photographs: Mark F Harris

Contents

Forward	*5*
Introduction	*7*
Dedication	*9*
The House Range	*11*
C. D. Walcott	*15*
Collecting and Studies Since 1950	*19*
About Trilobites	*21*
Cambrian Environment	*25*
Wheeler Amphitheater	*29*
Binomial Nomenclature	*37*
Robert Harris Collection	*39*
Rock Shop History	*81*
Success Story	*85*
Reviews	*89*
About the Author	*95*
Research Sources	*97*

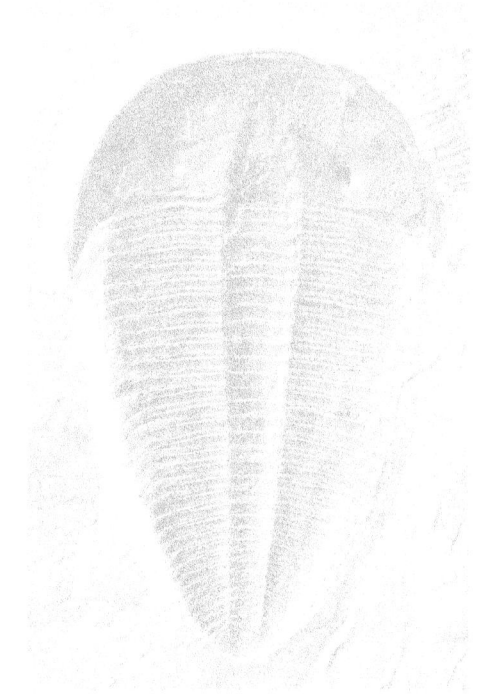

Forward

My childhood hometown was Delta, a rather isolated farm community in west central Utah on the eastern side of the great Pahvant Valley. My father was a schoolteacher and a farmer, but one of his greatest interests was geology. As a young man in the early 1900s, he worked in the Old Hickory Mine near Milford, Utah.

Extending some fifty miles beyond Delta is the "West Desert" where my father would take my brothers and me to mountains or hills by the names of Notch Peak, Swasey Mountain, Drum Mountain and Topaz Mountain, as well as to the old mining town of Joy. In these places he would collect topaz crystals, agate and jasper rocks and especially fossil trilobites.

Abundant fossil trilobites could be collected from the dry stream beds of the "Wheeler Amphitheater" below Swasey Peak of the House Range, where over long periods of time, the calcified trilobites were freed from weathered Wheeler Shale. I remember getting on my hands and knees, crawling down the small dry steam beds collecting fragments and often complete trilobite specimens. We also discovered that certain layers of the Wheeler Shale had a myriad of fossil trilobites. This permitted the collection of matrix specimens.

This early activity with our father caused me and my four brothers to have an interest in geology. My brother Robert was most interested in collecting trilobites and eventually made a career of marketing trilobite fossils and other geological materials. By far the most common and widely distributed trilobite found in the Wheeler Shale is ***Elrathia kingi***, but in the various formations of the House Range are numerous species. In this book are pictures and brief descriptions of the fossil trilobites Robert has collected over the many years.

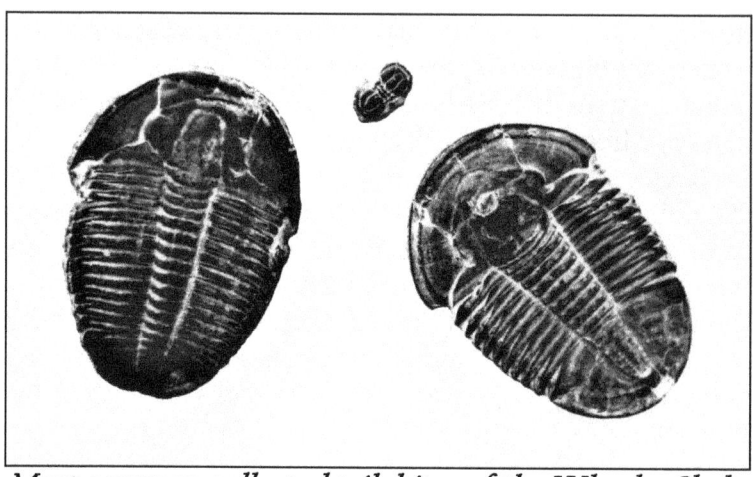

Most common collected trilobites of the Wheeler Shale: Elrathia kingi, Peronopsis interstrictus, Asaphiscus wheeleri

Introduction

Rather than being a scientific document this book is mainly a compilation of photographs of Utah trilobites. All trilobites pictured in this publication are from the three upper formations of the House Range: Wheeler Shale, Marjum Formation and Weeks Limestone.

The fossils in this Robert Harris Collection are some of the better-preserved trilobites found in these formations.

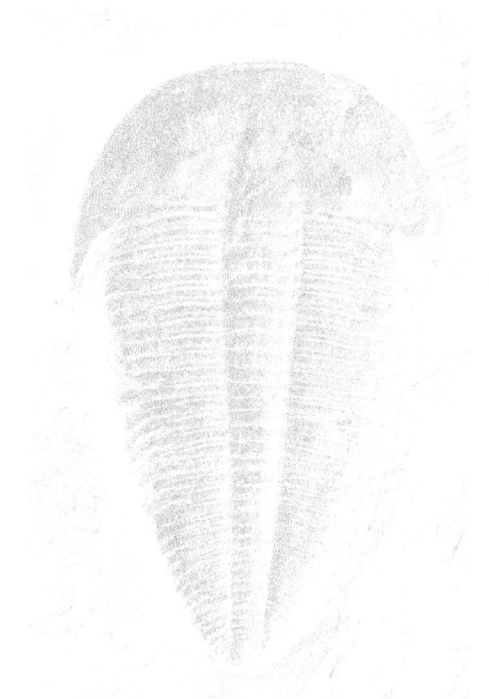

Dedication

This book is respectively dedicated to our father, Edward Daniel Harris (1894-1980), a man of many skills. By profession, he was a school administrator and teacher, but also a farmer and carpenter. He also served in various capacities in his church, both in teaching and leadership positions, and for three years as a young man was a missionary to the Maori people of New Zealand. He was a good public speaker and an entertainer, telling stories and jokes and performing sleight of hand tricks.

While working in the Old Hickory Mine near Milford, Utah to pay his college tuition, he gained a basic understanding of geology. After marrying and establishing a home in Delta, he often explored the mountains around the area and prospected for various minerals and fossils. This interest in geology was passed on to his children, some of whom have made professions in some aspect of the field. Robert attributes much of his success in the rock and fossil business to his father.

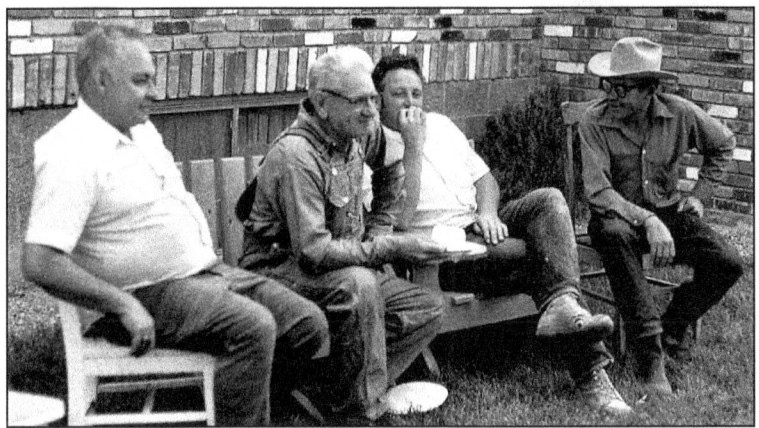

Edward Daniel Harris with his sons: Ted, Robert and John, all of whom related to some aspect of geology (Mark Harris Photo)

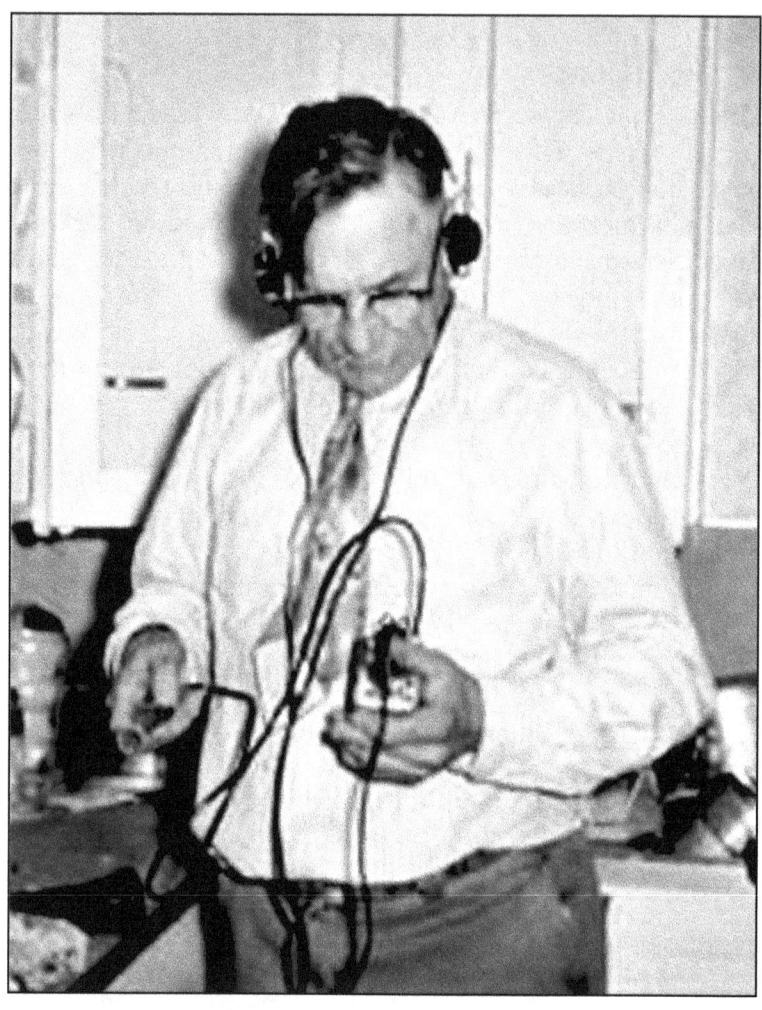

*Edward Daniel Harris with his Geiger Counter that he used to prospect for uranium in the West Desert Mountains during the 1970s
(Ted Harris Photo)*

The House Range

Swasey Peak and Notch Peak are the two main prominences of the House Range.

Swasey Peak
(https://ucmp.berkeley.edu/cambrian/house.html)

Western Utah is the home of one of the best-known Cambrian fossil localities in the world. The slopes of Swasey Peak in the House Range are composed of a rock layer known as the Wheeler Shale, with the overlying Marjum Formation forming the top of the peak. The Wheeler Shale and Marjum Formation, strata of Middle Cambrian age, are exposed throughout the House Range and nearby mountain ranges west of the town of Delta, Utah. The Wheeler Shale is named for a great bowl-shaped feature in the House Range known as the Wheeler Amphitheater, while the Marjum Formation is named for its outcrops at Marjum Pass, also in the House Range. Much of the Wheeler Shale is quite unfossiliferous, but certain layers contain abundant trilobites and other shelly fossils. The Wheeler Shale and Marjum Formation also contain a diverse biota of soft-bodied fossils, including many of the same taxa found in the more famous Burgess Shale.

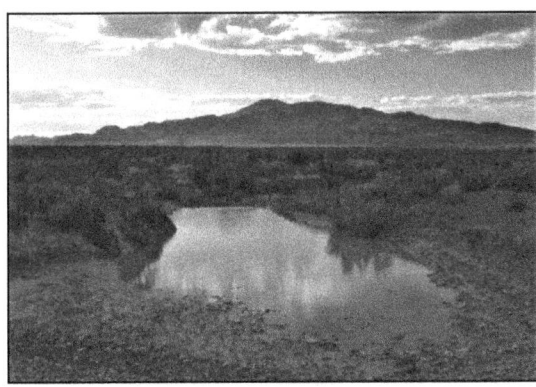

Swasey Peak

Photograph by Mark Harris

Notch Peak: Utah's Equivalent of El Capitan

(Tina Crowder: Community Columnist, April 11, 2014)

Hidden away in the House Range 50 miles west of Delta, nearly to the Nevada border, is one of the highest cliffs in North America with a vertical drop of 2,200 feet!

If you were to drive west from Delta on Highway 6, you might not even notice it was there; however, if you were to approach it from the west you wouldn't be able to take your eyes off this massive rock cliff known as Notch Peak.

Home to the 500-million-year-old trilobite fossil, the ancient bristlecone pine and fine gold dust deposits, Notch Peak boasts the second highest continuous vertical rock face in the nation behind Yosemite's El Capitan (at about 3,000 feet). It is considered by rock climbers to be one of the most difficult climbs in Utah. Rest assured you won't have to take up rock climbing and risk your life to get there, but can easily hike up to Notch Peak from Sawtooth Canyon, a 4.5-mile trail that climbs about 2,800 feet while it gradually winds up to the summit, for a 9-mile round trip adventure.

+++++++

Author's note: Within the Cambrian-age Wheeler Shale, Marjum Limestone and Weeks Formation of the House Range are found fossil trilobites that are known around the world, with Robert Harris being one of the prime marketers and promoters.

The geological formations can better be viewed in the western escarpment of Notch Peak than in Swasey Peak even though fossils are more accessible on Swasey Mountain.

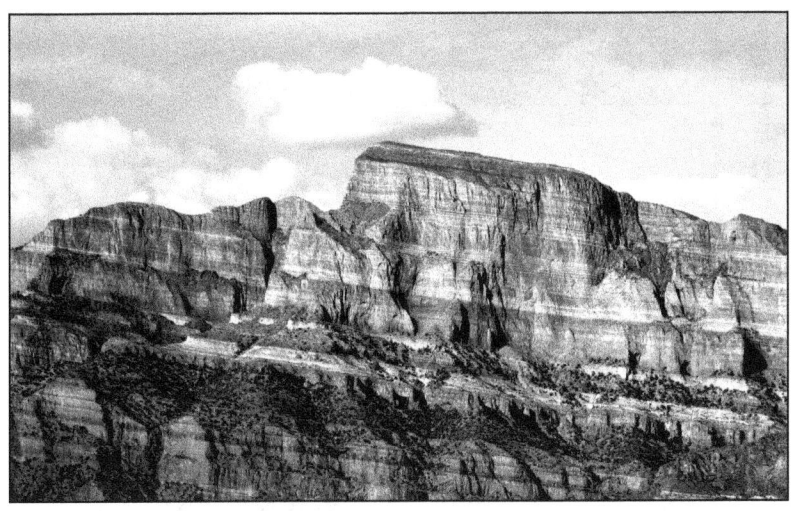

Notch Peak
(Photograph by Mark Harris)

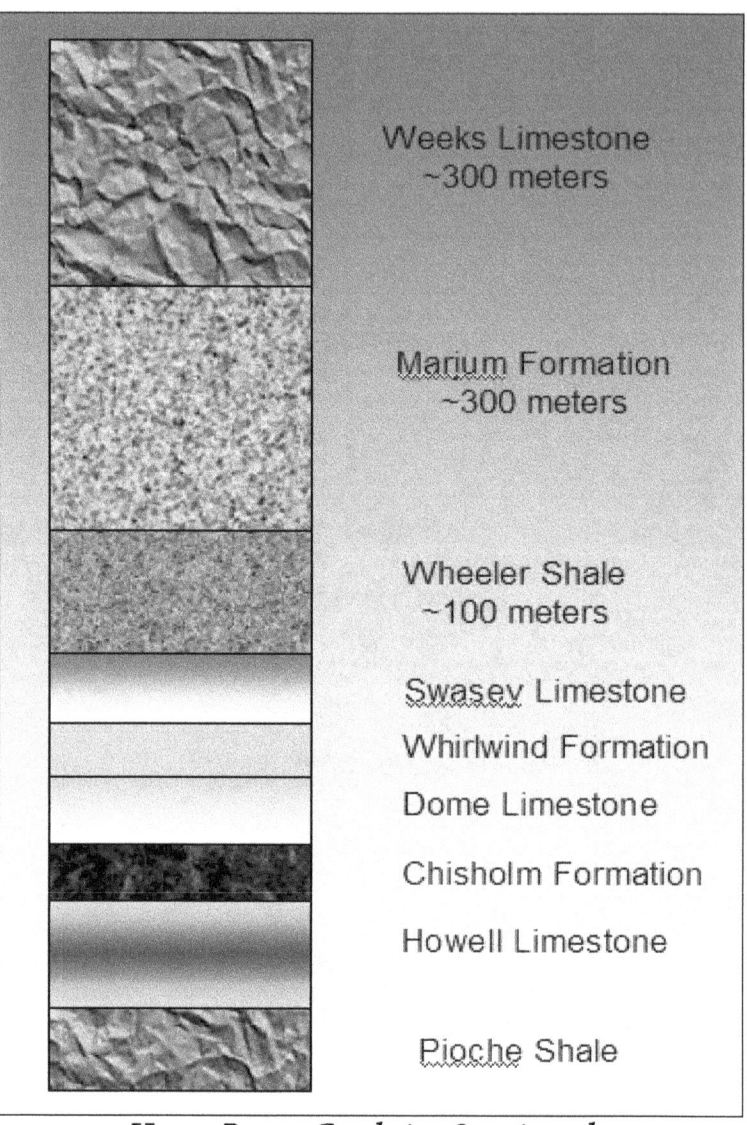

House Range Cambrian Stratigraphy

(From: Hintze and Robison, Middle Cambrian Stratigraphy of the House, Wah Wah, and adjacent ranges in Western Utah; Geological Society of American Bulletin 86-881-891)

C. D. Walcott and Millard County's Wheeler Shale

(From: Geology of Millard County, Utah by Lehi F. Hintze and Fitzhugh D. Davis, Bulletin 133, Utah Geological Survey, 2003. Posted on Great Basin Museum website)

There is a fascinating old photograph on page 28 of The Geology of Millard County, Utah by Hintze and Davis, titled "Charles D. Walcott's unpublished photograph of his paleontological exploration party in Tule Valley on September 11, 1903". We see a buckboard pulled by a two-horse team and a canvas covered supply wagon with a team of four. There are three carefully posed men and a dog. Water barrels are strapped to the wagon along with other provisions giving all indications of a well-supplied expedition.

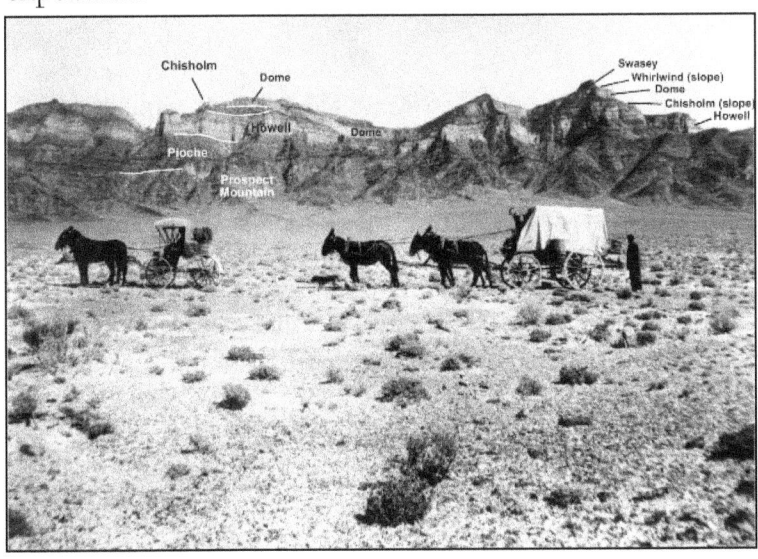

C. D. Walcott's 1903 exploration party in Tule Valley west of the House Range

C.D. Walcott was born with an insatiable interest in natural history. He was committed by the age of 17 to devote his life to the study of the oldest fossiliferous rocks of North America. At that time the oldest known were from the Cambrian of the western states and in particular the Wheeler Shale of Millard County. The Wheeler Geological Survey had stumbled upon this formation in the mid-1800s and a number of very interesting and primitive fossils had been sent back East for study and publication. The potential was certainly there, and Walcott was more than anxious to leave his office and conduct a systematic survey of these largely unexplored rocks.

He arrived at Salt Lake in August 1903 and left on the 27th with his field assistant Fred B. Weeks, his teamster Dan Orr, and cook Arthur Brown. It turned out to be a 26-day excursion into the desert wilderness of western Utah, most of that time within the confines of our County. The trip was so successful that he returned for a second field trip in 1905.

It is impossible to over-estimate the influence and prestige of Walcott: Hintze calls him "the father of Cambrian paleontology and stratigraphy in North America." At this particular time, he was director of the U.S. Geological Survey; a few years later he was president of the Smithsonian, president of the National Academy of Sciences, vice chairman of the National Research Council, chairman of the Carnegie Institute Executive Committee, and chairman of the National Advisory Committee for Aeronautics. Though much too busy for field work and exploration, our reading gives us a glimpse into the excitement Walcott experienced as he prepared for these western trips.

Walcott's 2nd U.S. Geological Survey Bulletin included the first good descriptions of five fossils represented in our present fossil collection. Their generic names are *Asaphiscus, Ptychoparia, Agnostus, Olenoides*, and *Acrothele.*

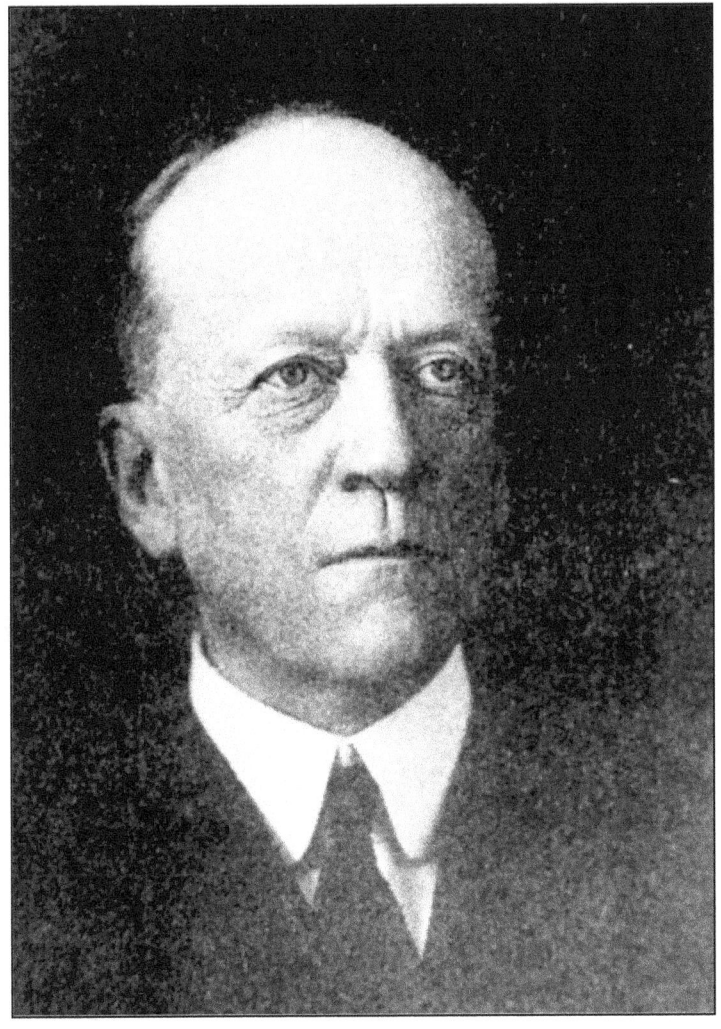

C. D. Walcott

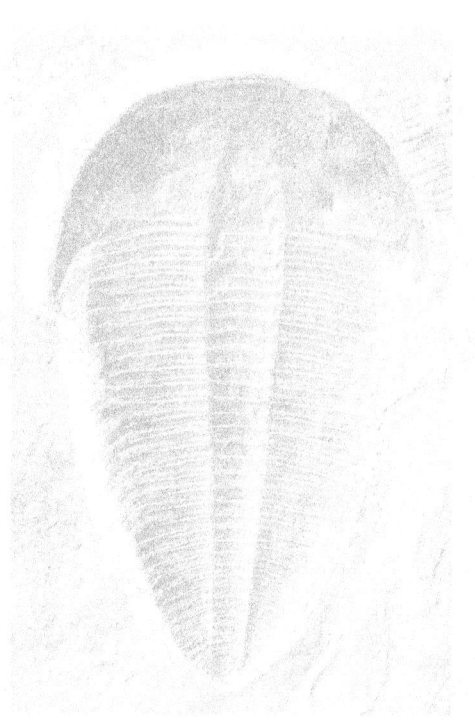

Collecting and Study Since 1950

*(From the Publication:
Exceptional Cambrian Fossils from Utah
A Window into The Age of Trilobites
by Richard A Robison, Loren E Babcock and
Val G Gunther, p 11-12)*

Interest in the Cambrian rocks and fossils of Utah substantially increased after World War II. Varied investigations by geologists and paleontologists of state and federal agencies, oil and gas companies, and mining companies produced important new information. A.R. (Pete) Palmer (1960, 1965a, 1965b), Cambrian paleontologists of the U.S. Geological Survey, first drew attention to distinct Cambrian sediment belts surrounding Laurentia, pioneered innovative biostratigraphic studies, and discovered evidence of significant trilobite extinction events.

The Cambrian of Utah also became a popular academic training ground. Since 1950, increasing numbers of faculty and students from many universities produced an even greater flow of new information about the Cambrian of Utah. Much of that information has been published, but a significant part remains in unpublished student works.

A knowledge of Cambrian fossils from Utah increased in recent decades, so too has interest in their acquisition for pleasure by amateur collectors and for profit by commercial collectors. The House Range, with its thick Cambrian section and several fossiliferous formations, has been a magnet for collectors. A day seldom passes without fossil hunters on its Cambrian outcrops. Less accessible localities in the Drum Mountains of western Utah and the Wellsville Mountains in northern Utah are runners-up.

Amateur fossil collectors have made many significant contributions to science, commonly by donating rare or unusual discoveries for scientific study and deposit in public museums. Such fossils become part of our natural heritage and remain available to the world for further investigation. No amateur fossil collectors of Utah have contributed more to science than the late Lloyd Gunther and his family of Brigham City, Utah. They freely shared knowledge gained over many years and donated thousands of fossils to educational institutions and public museums, including many illustrated here. In 1984, for outstanding achievement in paleontology by amateurs, Lloyd, his late wife Metta, and son Val Gunther received the first Strimple Award from the Paleontological Society, the world's largest organization of paleontologists.

Commercial collecting of Cambrian fossils from state-owned land in Utah is legal only by permit. Robert Harris of Delta, Utah was the first to obtain a permit in the late 1960s. Subsequently, he and other permit holders have mined hundreds of thousands of trilobites, mostly *Elrathia kingii*, from the upper Wheeler Formation in the House Range. These have been widely marketed and have made *E. kingii* possibly the world's most familiar species of invertebrate fossil. *E. kingii* was nominated to become the State Fossil of Utah but lost a 1988 vote to the Jurassic dinosaur *Allosaurus*.

About Trilobites

(newworldencyclopedia.org)

Trilobites are hard-shelled, segmented members of the phylum Arthropoda and the class Trilobita that appear in the fossil record for almost 300 million years—from about 540 to 251 million years ago. They existed throughout almost all of the Paleozoic era, flourishing in the earlier part of it and slowly declining in the later part, finally going extinct in the Permian-Triassic extinction event about 251 million years ago.

The most common trilobites were about 20-70 mm (1-3.5 inches) in length, but over their long history they ranged in size from 1 mm-720 mm (.04-28 inches) and exhibited so much variation on their basic body plan that they are classified into nine (or possibly ten) orders with more than 15,000 species. The smallest species are presumed to have been part of the free-floating plankton, while the more common, mid-sized species probably walked along the sea floor filtering mud to obtain food, and the larger varieties may have been swimming predators. The trilobites are considered to be the first animals to have evolved true eyes.

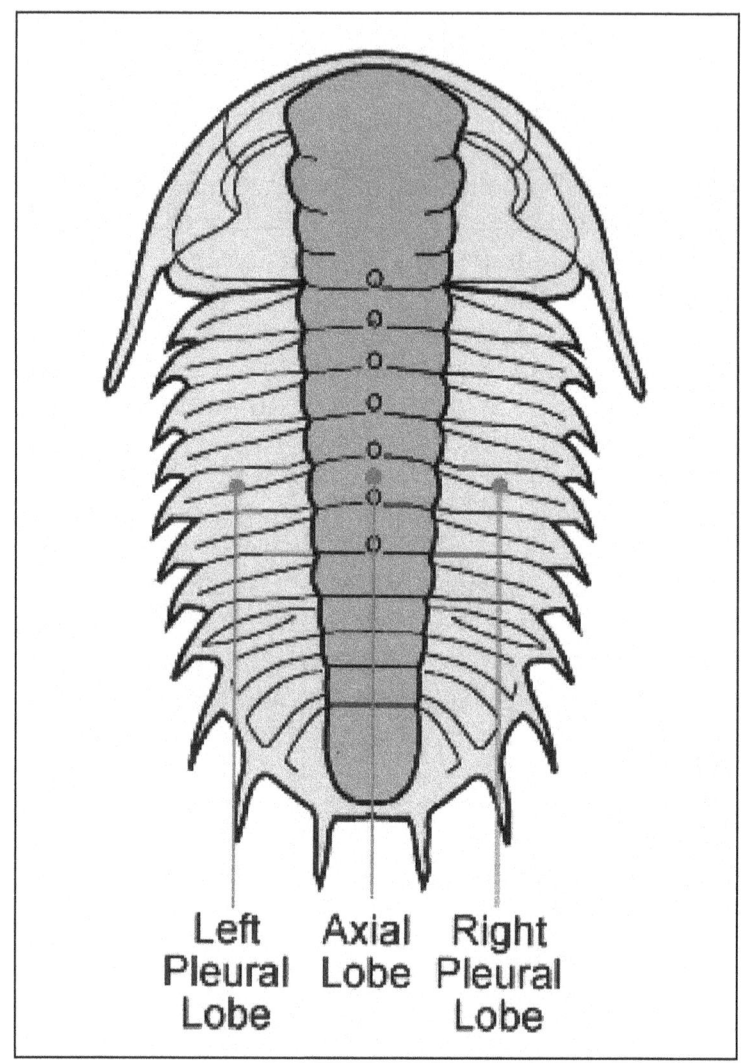

The name "trilobite" (meaning "three-lobed") is based on its three longitudinal lobes: The left pleural lobe, the central axial lobe, and the right pleural lobe.

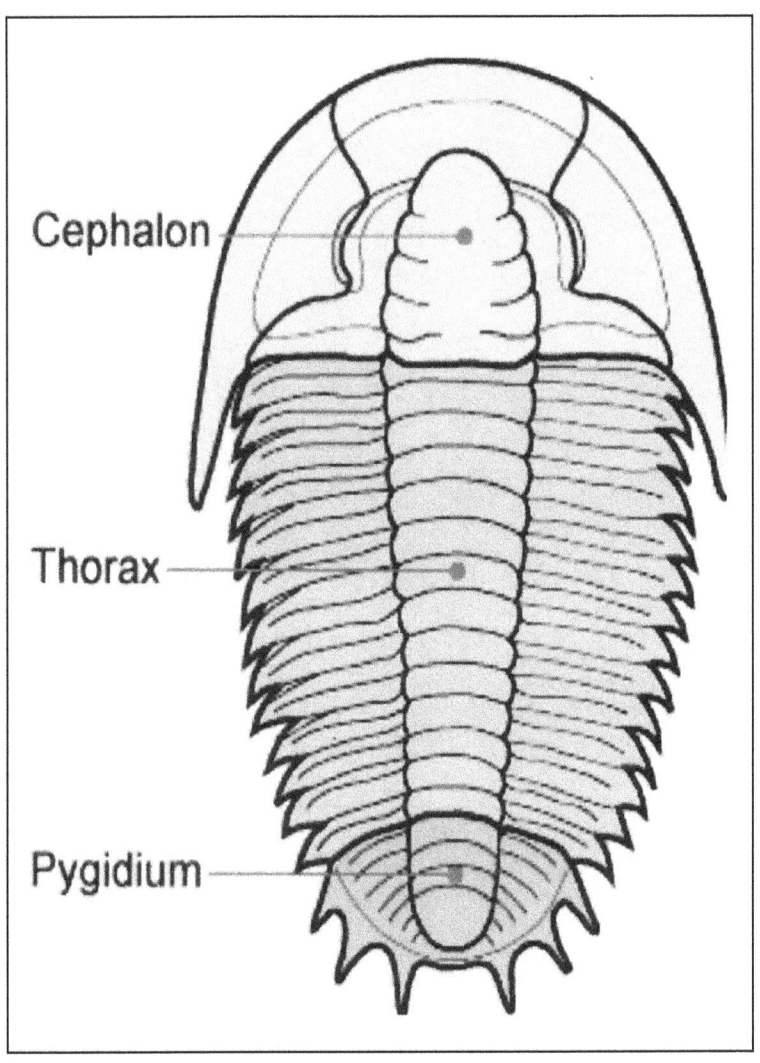

The trilobite body is divided into three major sections, a cephalon with eyes, mouthparts, and sensory organs such as antennae; a thorax of multiple similar segments (that in some species allowed the organism to curl up); and a pygidium, or tail section.

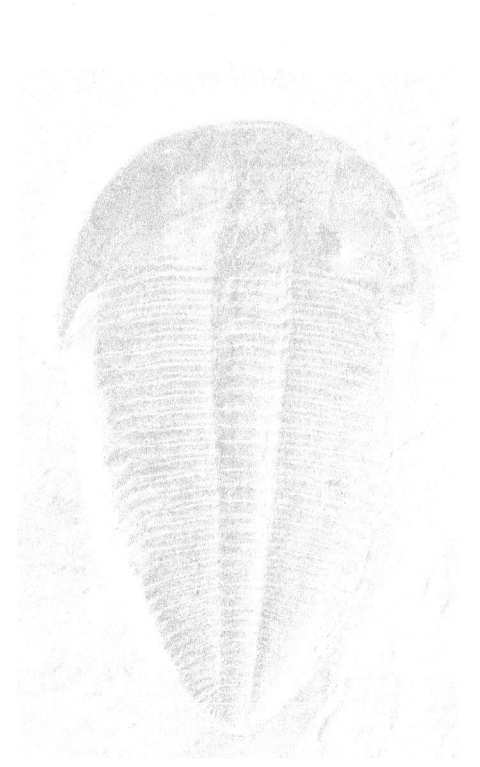

Trilobites and the Cambrian Environment of Utah

A Utah Geological Survey Publication
By Rebecca L. Hylland

Trilobites. The very name conjures up images from "B" science-fiction movies of bug-eyed, wiggly-legged, insect-like creatures that eat New York. Two questions we commonly receive are "what are trilobites and where are they found in Utah?"

What are Trilobites?

Trilobites are members of the phylum Arthropoda (jointed-foot animals). Arthropods have segmented bodies and appendages covered by an exoskeleton which provides support and protection for muscles and organs. Living Arthropods include insects, spiders, scorpions, ticks, crabs, lobsters, barnacles, and centipedes.

Trilobites belong to an extinct class of marine organisms called Trilobita. This name refers to the three-part (tri-lobes) latitudinal and longitudinal shape of a trilobite's exoskeleton. The latitudinal lobes consist of the cephalon (head), segmented thorax (body), and pygidium (tail); the longitudinal lobes consist of two lateral lobes (on each side of the body) and an axial lobe (central back area of the exoskeleton).

When did they live?

More than 500 different trilobite species have been found across Utah, in a broken band of Cambrian Period (570 to 500 million years old) limestones, siltstones, and shales that trends northeast-southwest across the western part of the state.

During the Early Cambrian (about 570 to 540 million years ago), western Utah was covered by a shallow sea. Slow-moving rivers flowed across the sandy lowlands of eastern Utah deposited sediment into the sea.

The heavier sediment (mostly sand) was deposited near the shoreline which metamorphosed through time into quartzite. The lighter sediments (mostly silt) were deposited farther out into the sea, and through time lithified into siltstone and shale. The deepest part of the sea was an ideal environment for the precipitation of calcium carbonate, which lithified to limestone.

Regional subsidence during the Middle and Late Cambrian (about 540 to 500 million years ago), caused the sea's shoreline to migrate eastward across Utah, allowing the deposition of a fairly complete sequence of Cambrian sediment in western Utah. Utah was located near the equator during the Cambrian, so the water temperature was warm.

The combination of warm, shallow water and nutrient-rich silt allowed several marine genera to thrive. The most common and diverse of these were trilobites, which occupied several different marine environments.

Where did they live in the sea?
Most trilobite species were bottom dwellers that crawled over sand and mud. Some of them could curl up like modern pill bugs. Other trilobites burrowed into bottom sand and mud using their shovel-shaped cephalons.

These crawling and burrowing trilobites were either scavengers, or they ingested mud and silt, digesting the organic material contained in it like modern day worms (annelids). Some trilobites lived in shallow burrows where they could keep their heads near the surface of the sand or mud, and grab passing prey.

Fossil evidence suggests some trilobites were capable of swimming. The bodies of swimming trilobites are narrower, and the eyes are closer to the sides of the cephalon, than those of bottom-dwelling trilobites. Swimming trilobites may have been predators, or they may have been "filter-feeders" using special appendages to remove nutrients from the surrounding water. The smallest trilobites were plankton-like and lived close to the water surface.

Where are trilobites found in Utah?
Trilobites are probably the most common fossils collected in Utah, many world-class specimens from this state reside in museums throughout the world. In Utah, trilobites can be found at several localities.

House Range
The Wheeler Amphitheater in the House Range, Millard County is one of the more well-known collecting areas. Most of the trilobites in this area come from the Middle Cambrian formation called the Wheeler Shale. The Wheeler Shale contains interbeds of shaley limestone, mudstone, and thin platy limestone. Another trilobite-bearing unit that directly overlies the Wheeler Shale in the central part of the House Range is the Marjum Formation. This formation consists of thin-bedded, fine-grained, silty limestone with interbeds of shale and mudstone.

Also located in the central part of the House Range is a fossiliferous limestone called the Weeks Formation, that crops out in North Canyon near Notch Peak. The Weeks Formation overlies (is younger than) the Marjum Formation and also contains trilobites.

Wellsville Mountains
Another trilobite-bearing unit is the Spence Shale Member of the Langston Formation in the Wellsville Mountains, Box Elder County. Here, trilobites can be found in Miner's Hollow, Cataract Canyon, Dry Canyon, and the area between Antimony and Hanson Canyons.

Wheeler Amphitheater
(http://www.trilobites.info/Utah.htm)

Some Excerpts: The Wheeler Amphitheater, also known as Antelope Springs, is a world-famous fossil collecting location for trilobites.

The trilobites often weather out of the rock and can be found loose on exposed slopes. The two most common large trilobites are *Asaphiscus wheeleri* (abt 2 inches) and *Elrathia kingi* (abt 1.6 inches). The most common smaller trilobites in this area are *Peronopsis interstrictus* and *Hypagnostus parvifrons*, both about 1/4 inch long.

How to get there: From Delta, travel 32 miles west on Highway 6/50. At the Long Ridge Reservoir sign between mile markers 56-57, turn right. Then travel roughly 20 miles down a gravel road to reach the trilobite area.

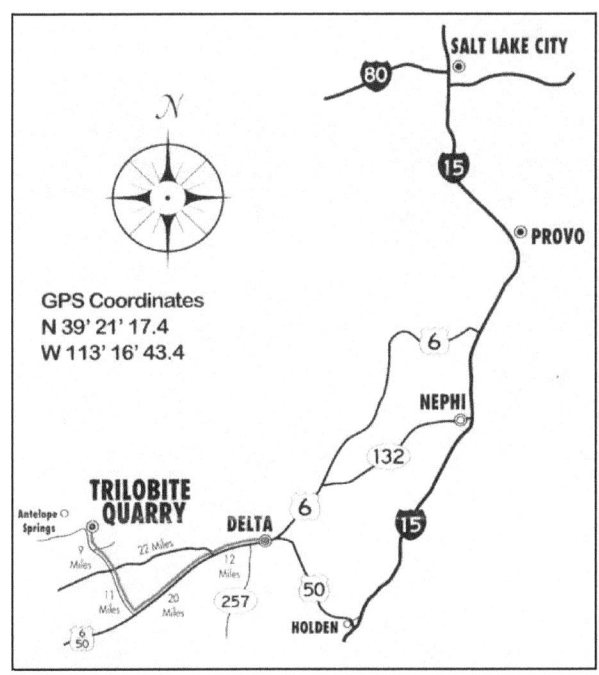

U-DIG FOSSIL MAP

Where to collect: Trilobites can be found in many of the different layers of shale in Wheeler Amphitheater.

There are three techniques to find trilobites. The easiest is to pay at one of the nearby quarries. They have tools and everything else you may need.

The moderately difficult way is to find where someone has split the shale rocks to search for the hidden trilobites and search from there.

The most difficult technique is to wander along the dry washes until you find an outcrop of rock. At that point, start to dig and split it into layers, being sure to look on both sides for trilobites or imprints.

The trilobites vary in size from as small as 1/4 inch to 2 inches long. There are several different species, so if you find several, you may notice differences.

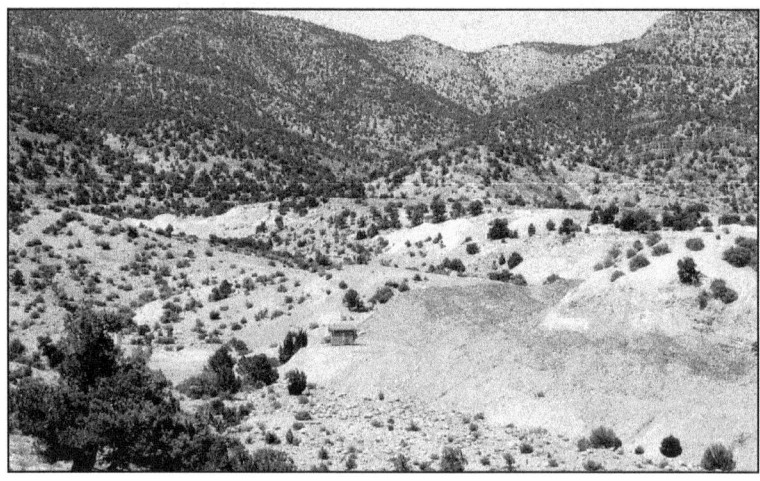

Wheeler Amphitheater Fossil Site
(Photo from U-DIG FOSSIL Website)

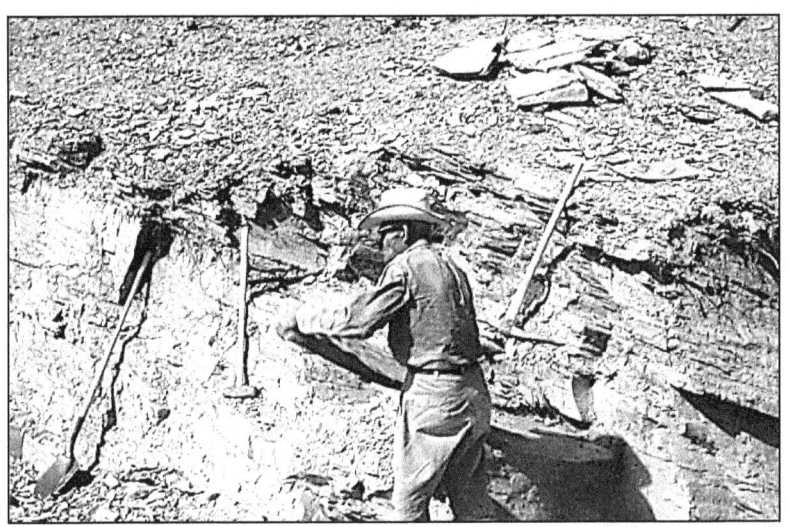

Robert's Uncle Frank Forster working in a Wheeler Shale quarry (Mark Harris Photo)

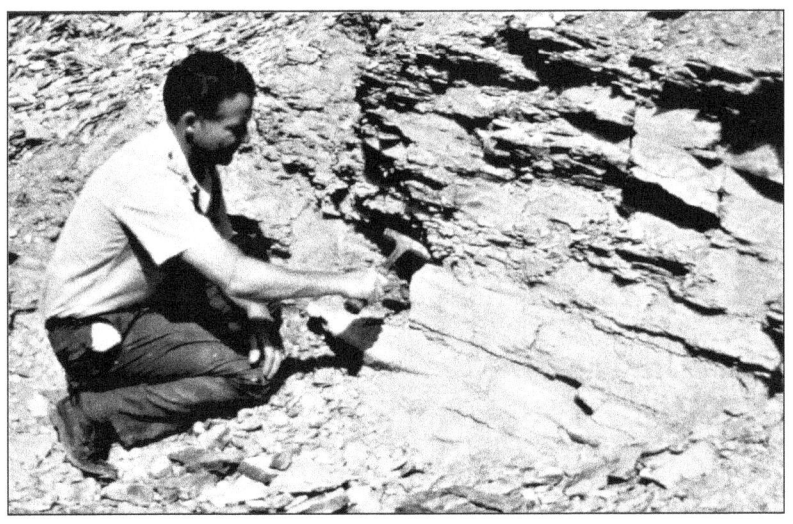

Mark Harris working in a Wheeler Shale quarry

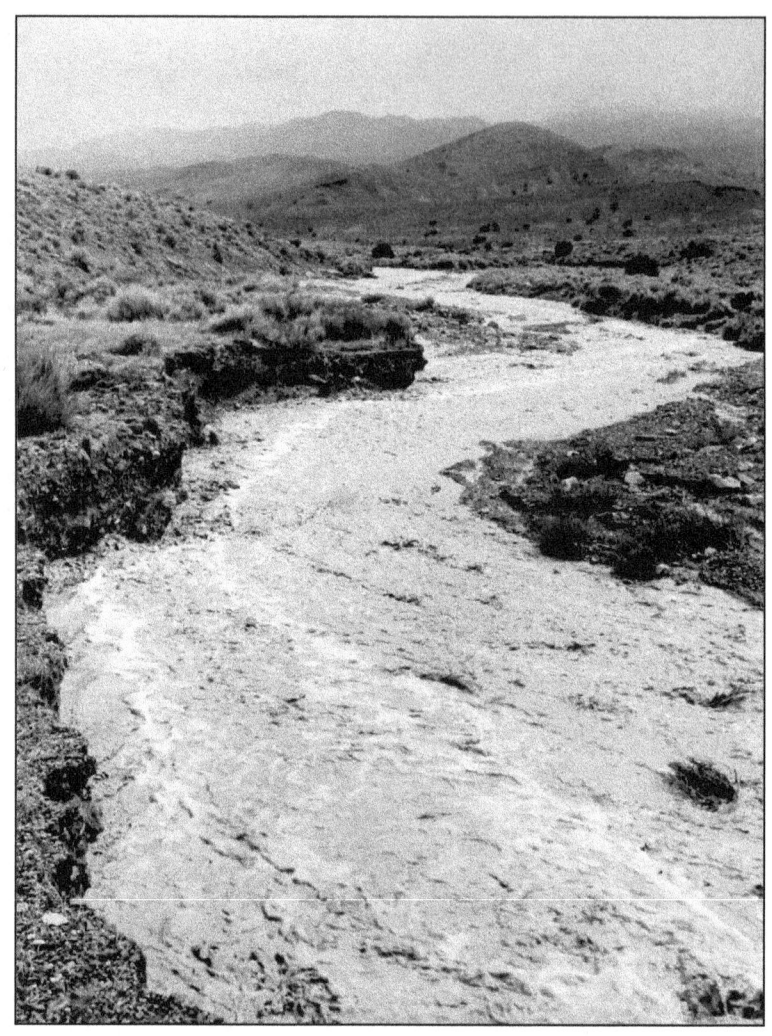

Small streambeds at Antelope Springs are usually dry, but after flash floods from rare summer thunder showers the eroded shale fragments are "stirred up" a bit making it easier to find trilobites (Mark Harris Photo)

Excerpts and Sketches Regarding Wheeler Amphitheater

From "Geologic Road Guides of Western Utah and Eastern Nevada" by Lehi F. Hintze-Department of Geology, Brigham Young University, p 27-29, published in 1973

Trilobites occur in many of the horizons in the Wheeler Shale and lower Marjum Formation, but in most places the trilobites cannot be liberated from the matrix as easily as they can here, The trilobites here are not silicified but have been thickened by a "fortification structure" of calcite, which has grown between the top and the bottom of the trilobite carapace, strengthening it so that the trilobites can be broken entirely free of the matrix.

Trilobites from here were first collected by Indians who used them as charms. Charles D. Walcott, the Director of the U.S. Geological Survey, spent the entire summer of 1905 collecting Cambrian fossils in the House Range, naming about 50 new species of trilobites taken mostly from the Wheeler and Marjum formations. A detailed restudy and careful zonation of the trilobites of the Wheeler and Marjum formations are published by Robison (1964).

By far the most common trilobite in the Wheeler Shale is *Elrathia kingii;* the next two most common are the small agnostid *Peronopsis*, and the *Asaphiscus wheeleri*, which is characterized by its large pygidium (tail). Small phosphatic brachiopods are also common. The horizon from which whole trilobites are most commonly obtained is near the top of the Wheeler Shale and is marked by reject talus for the "strip mining" operation now necessary to obtain many good specimens. This is probably the most famous trilobite locality in the world. *Elrathia* specimens from here are found in every university and museum collection, and nearly every day amateur collectors are here digging. Robert Harris of Delta has been collecting trilobites commercially here for the past few years. He has turned the less common specimens over to Dr. Richard Robison of the University of Utah, who has described the recent finds (Robison, 1971). Fourteen different species of trilobites have been recovered so far from the Wheeler Shale, as have three species of phosphatic brachiopods, three sponges, and an early crinoid.

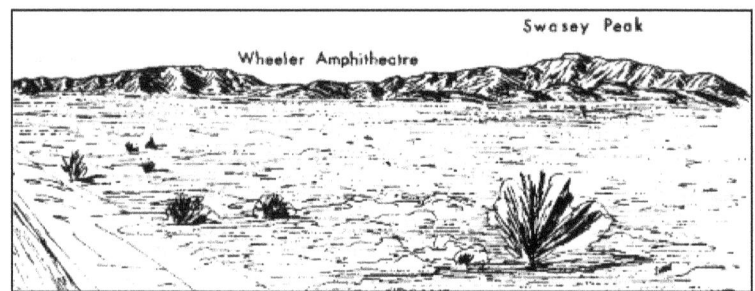

Sketch of House Range

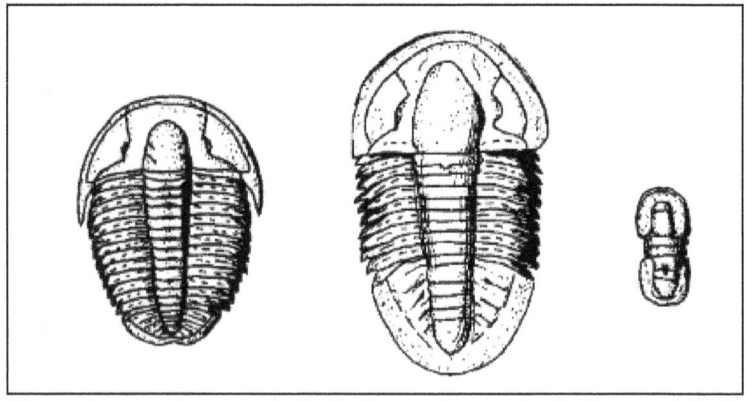

Three Most Common Trilobites from the Wheeler Shale Elrathia kingii, Asaphiscus Wheeleri, Peronopsis interstrictus

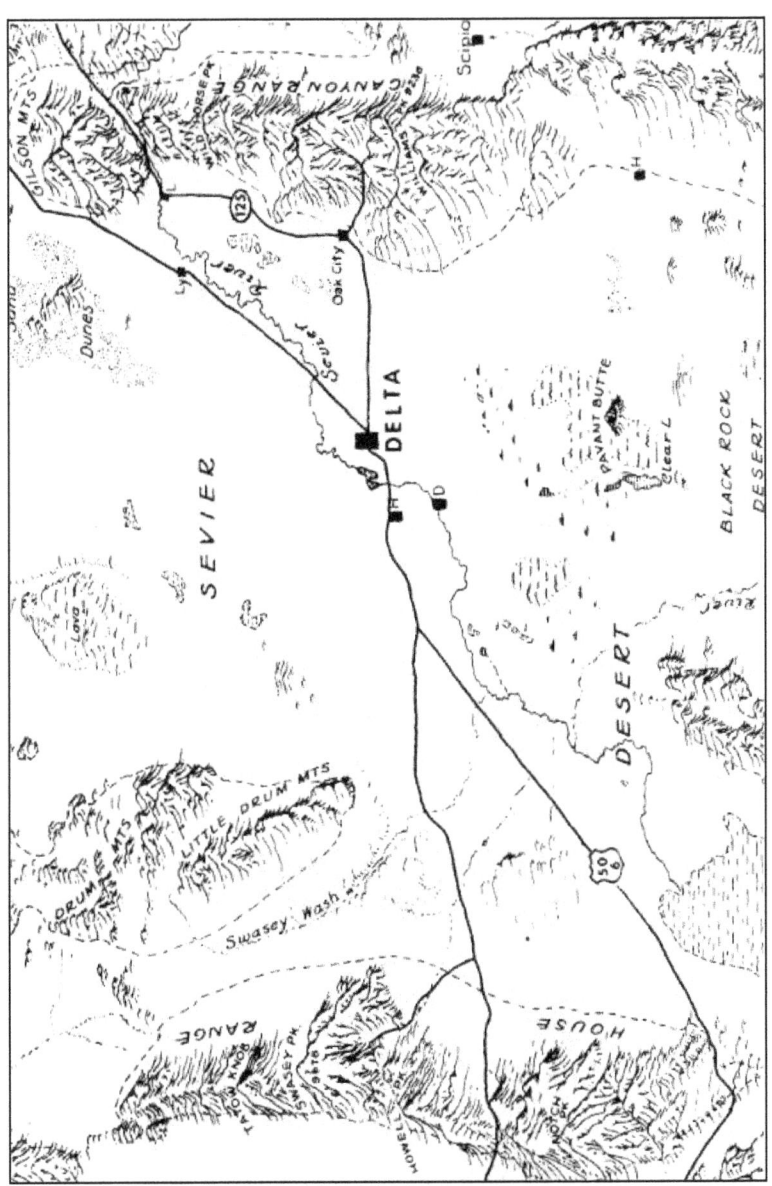
Hintze Map, Delta to House Range, 1973

UTAH TRILOBITES

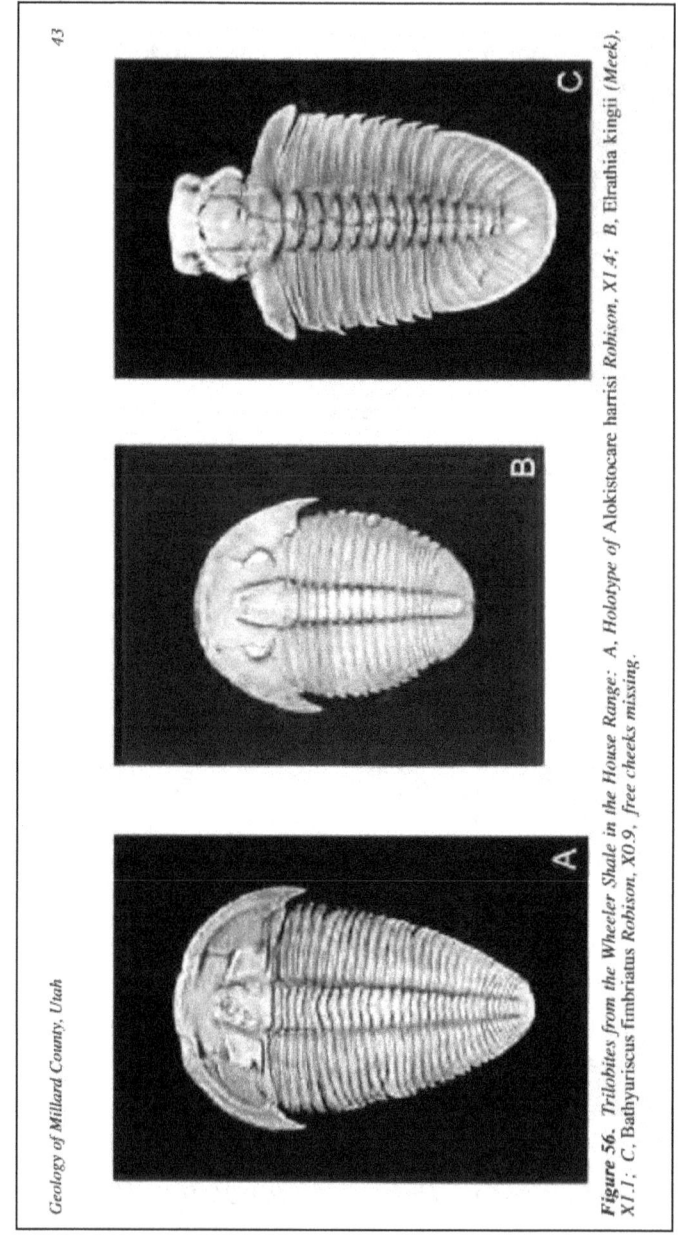

Figure 56. *Trilobites from the Wheeler Shale in the House Range: A, Holotype of Alokistocare harrisi Robison, X1.4; B, Elrathia kingii (Meek), X1.1; C, Bathyuriscus fimbriatus Robison, X0.9, free cheeks missing.*

From a 2003 publication by Hintze are shown three unique fossil trilobites found in the Wheeler Shale

Binomial Nomenclature

(http://www2.nau.edu/~bio372)

The present system of binomial nomenclature identifies each species by a scientific name of two words, Latin in form and usually derived from Greek or Latin roots. The first name (capitalized) is the genus of the organism, the second (not capitalized) is its species. The scientific name of the white oak is *Quercus alba*, while red oak is *Quercus rubra*. The first name applies to all species of the genus—Quercus is the name of all oaks—but the entire binomial applies only to a single species. Many scientific names describe some characteristic of the organism (alba=white; rubra=red); many are derived from the name of the discoverer or the geographic location of the organism. Genus and species names are always italicized when printed; the names of other taxa (families, etc.) are not. When a species (or several species of the same genus) is mentioned repeatedly, the genus may be abbreviated after its first mention, as in *Q. alba*. Subspecies are indicated by a trinomial; for example, the southern bald eagle is *Haliaeetus leucocephalus leucocephalus*, as distinguished from the northern bald eagle, *H. leucocephalus washingtoniensis*.

The advantages of scientific over common names are that they are accepted by speakers of all languages, that each name applies only to one species, and that each species has only one name. This avoids the confusion that often arises from the use of a common name to designate different things in different places (for example, see elk), or from the existence of several common names for a single species. There are two international organizations for the determination of the rules of nomenclature and the recording of specific names, one for zoology and one for botany. According to the rules they have established, the first name to be published (from the work of Linnaeus on) is the correct name of any organism unless it is reclassified in such a way as to affect that name (for example, if it is moved from one genus to another). In such a case definite rules of priority also apply.

+++++++

Author's note: What is a species? The general definition of a species is "a group of organisms that look more or less alike and in nature breed freely and have fertile offspring". This definition quite well applies to living organisms, both plants and animals where the offspring of the reproductive process can be observed. However, it's also observed that two closely related species can have offspring which are called hybrids, as in the mating of a donkey with a horse, the offspring being a mule. But in most cases the mule is sterile. Anther observed example is the mating of a wolf with a domesticated dog, in which case many of the offspring are fertile. So, the binomial system of naming, although not perfect, is very helpful in identifying and classifying organisms.

This two-name system of identification is also applied to trilobites and other extinct organisms found only as fossils. Here, the similarities, or differences in physical traits are key to the naming process.

Robert Harris Collection of Trilobites

On the following pages are photographs of Utah trilobites collected by Robert Harris. Below each picture is the scientific name (genus and species), the date and by whom the specimen was identified or described, the geological formation where it was discovered and its relative size.

Carlolus Linnaeus (1707-1778), was a Swedish botanist, physician, and zoologist who formalized binomial nomenclature, the modern system of naming organisms. He is known as the "father of modern taxonomy". During his lifetime he classified over 10,000 species, both plants and animals. This included fossils of extinct organisms. In 1757 Linneaus identified the trilobite *Agnostus pisiformis*.

In the mid-1800s many new species of trilobites were identified, with each being given a genus and species name. Research continues today. As additional information is discovered about trilobites, the naming process is refined, where, in many cases, a trilobite is given a different name, often the genus name. Also, scientific literature shows that in some cases, there is insufficient knowledge regarding the date and the author who first or best described a species. For this publication, two books: *The Back to the Past Museum Guide to Trilobites* by Enrico Bonino and Carlo Kier and *Exceptional Cambrian Fossils from Utah, A Window into The Age of Trilobites* by Richard A Robison, Loren E Babcock and Val G Gunther and also the *Western Trilobite* website will be the sources.

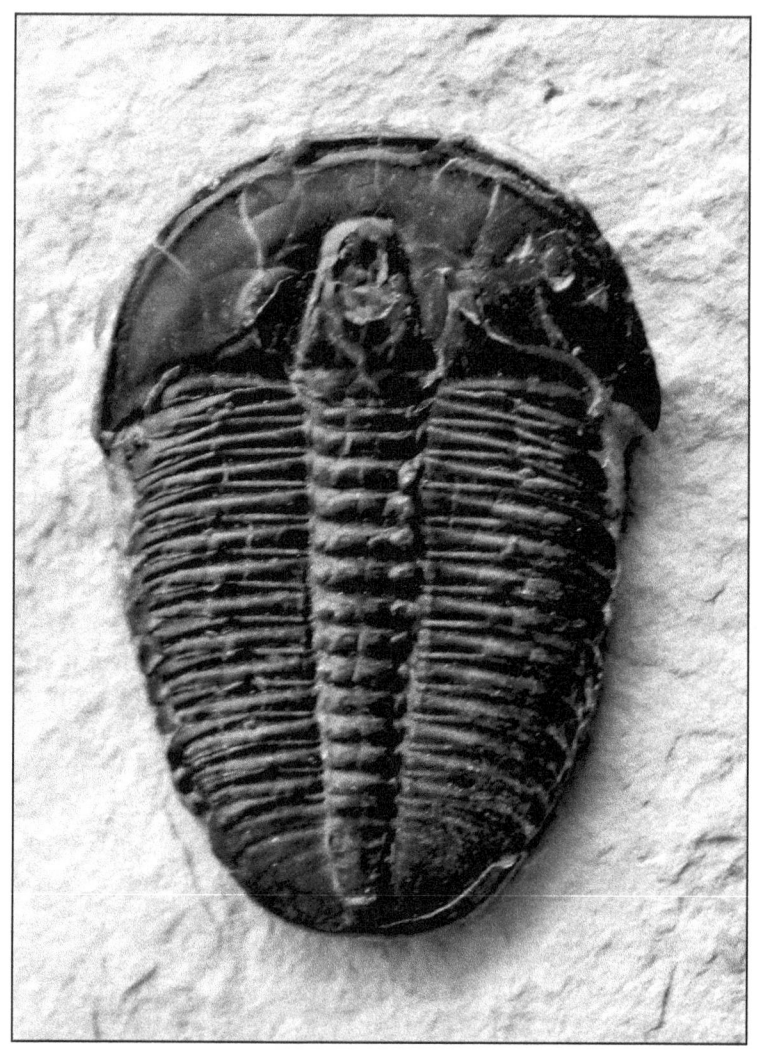

Elrathia kingi (Meek, 1870)
Wheeler Formation
America's most collected trilobite
Length: up to 50 mm

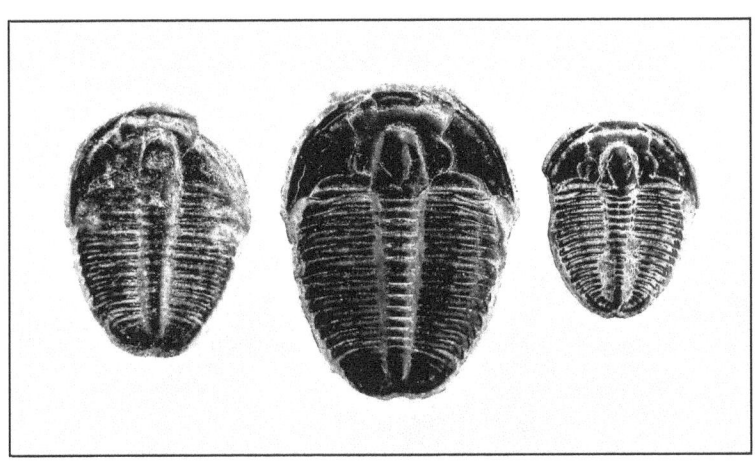

Elrathia kingi trilobites of varying size

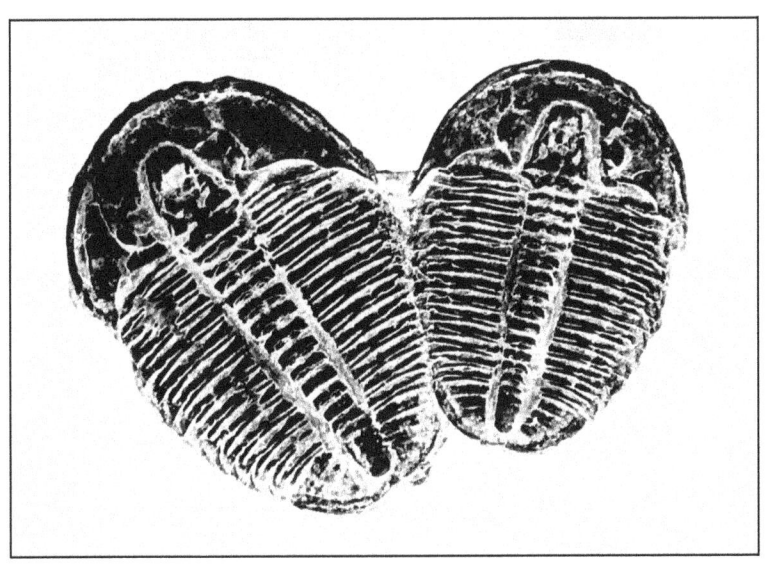

Elrathi kingi trilobites attached

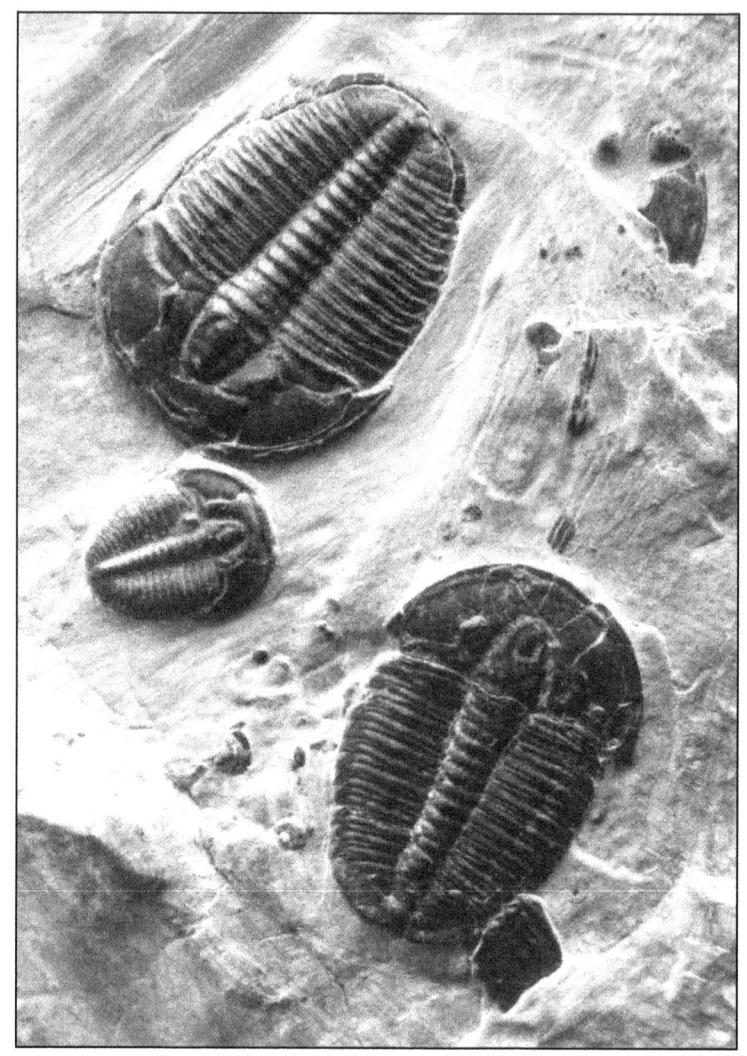

Elrathia kingi trilobites, matrix specimens

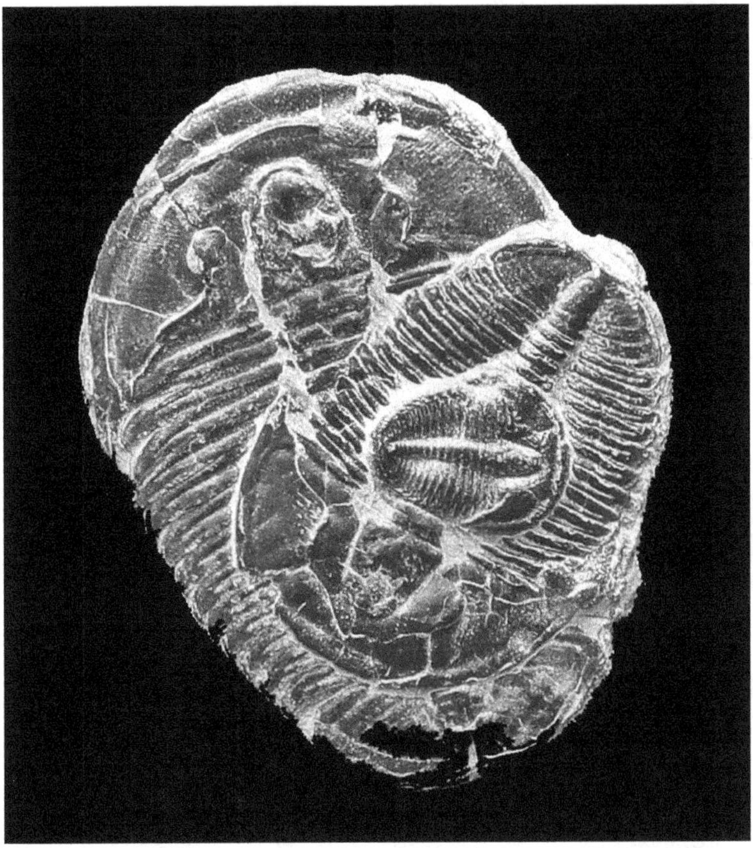

A cluster of Elrathia kingi trilobites

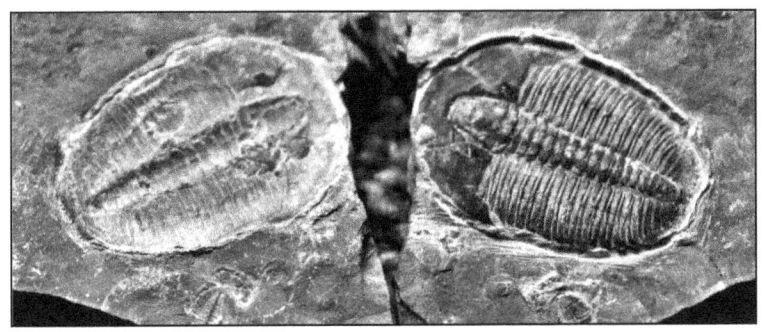

Elrathia kingi matrix specimen and "cast"

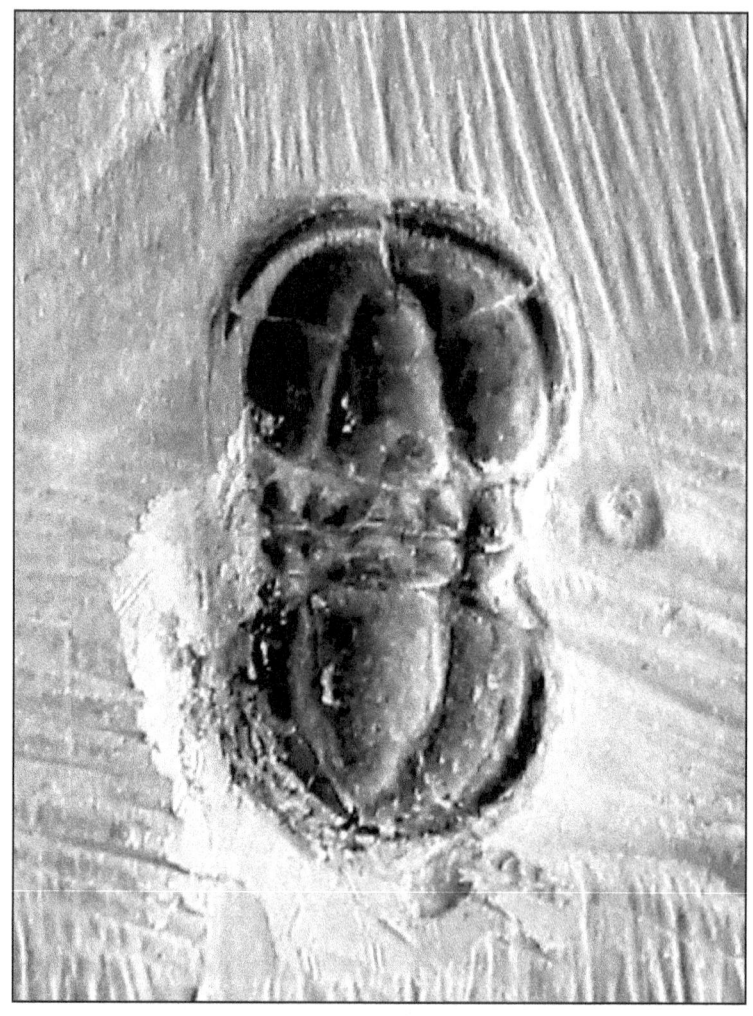

Peronopsis interstrictus (White, 1874)
Wheeler Shale
One of the world's smallest trilobites
Length: up to 9 mm

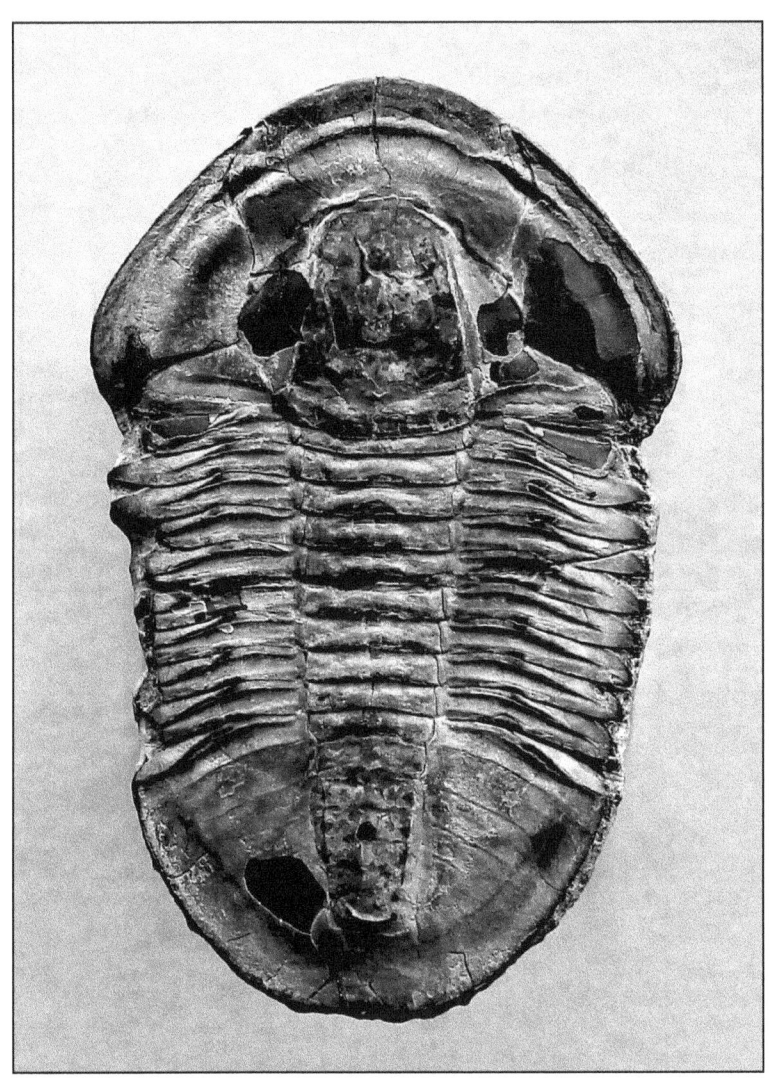

Asaphiscus wheeleri (Meek, 1873)
Wheeler Shale
Length: up to 100 mm
This large specimen discovered by John F. Harris

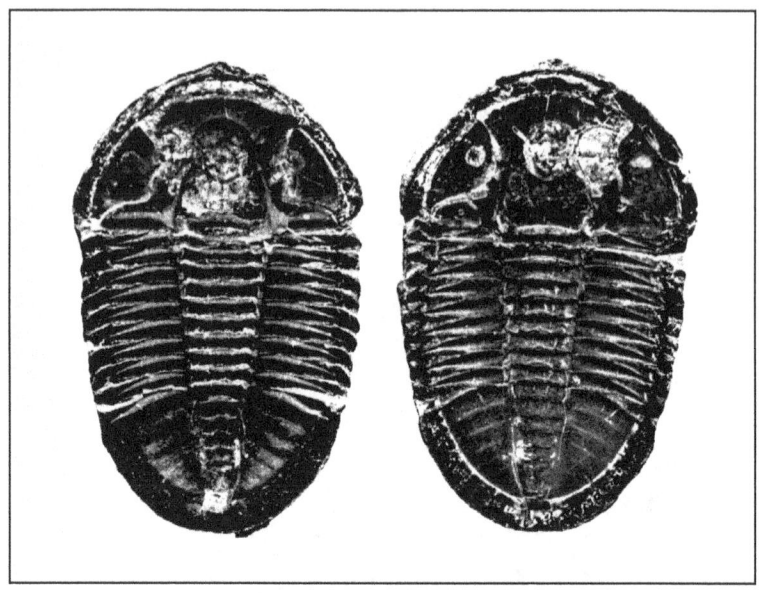

Asaphiscus wheeleri
Left: Dorsal surface   Right: Impression of dorsal surface

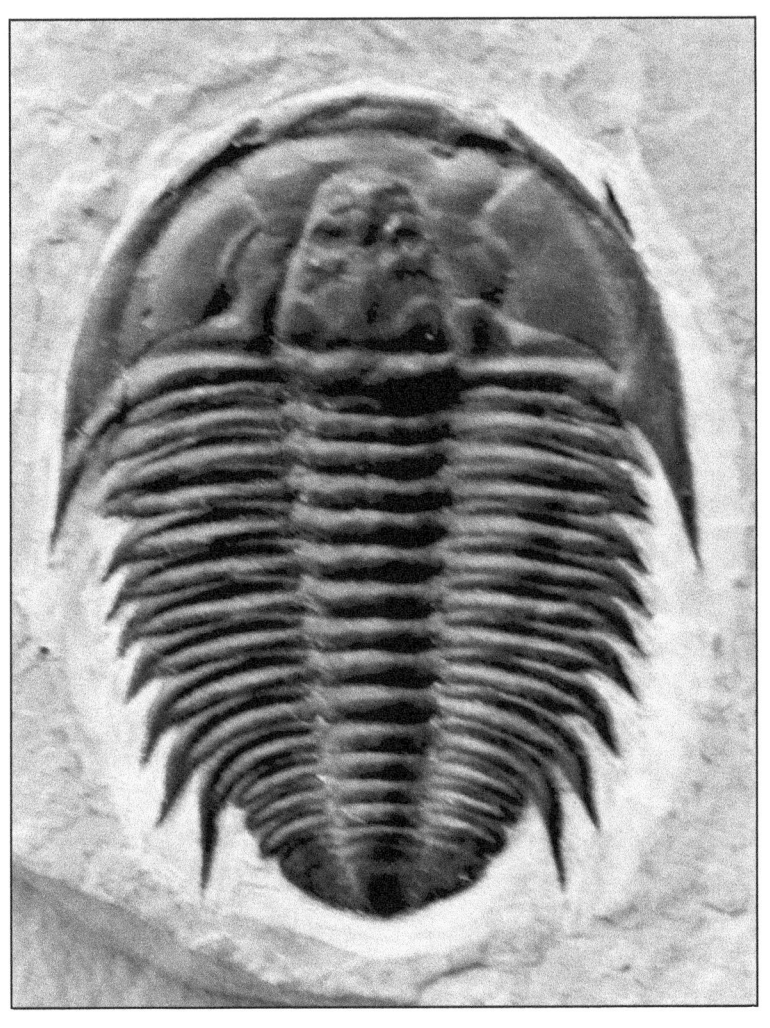

Modocia typicalis (Resser, 1938)
Marjum Formation
Length: about 20 mm

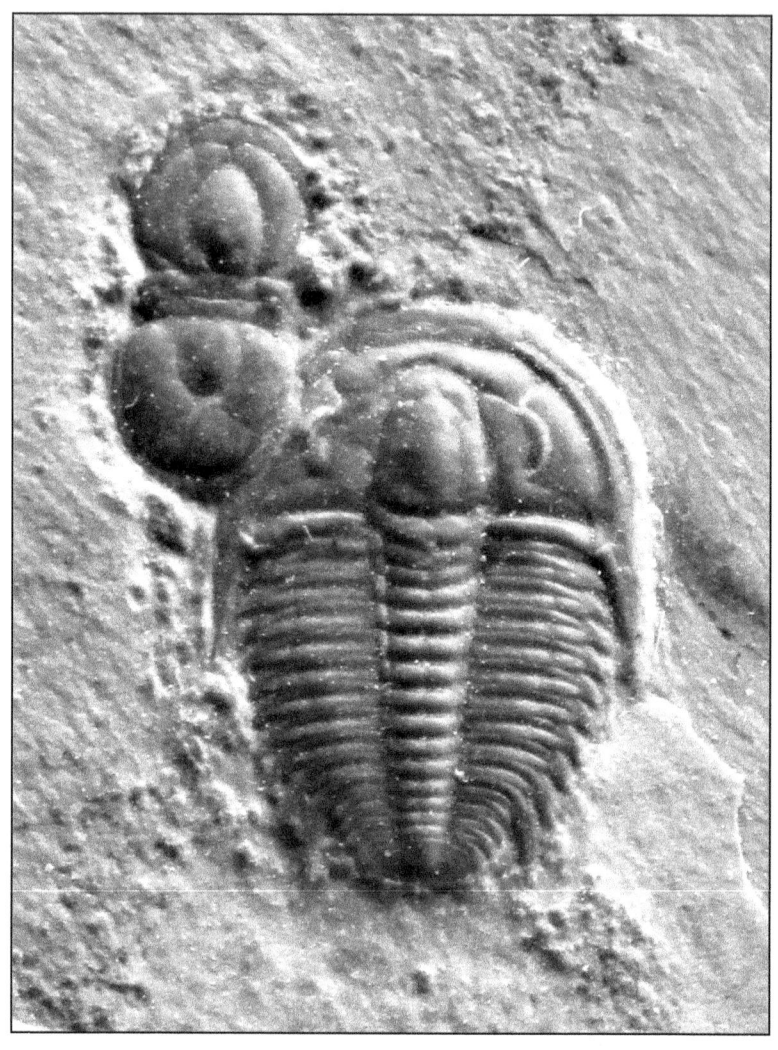

Peronopsis interstrictus and Modocia typicalis

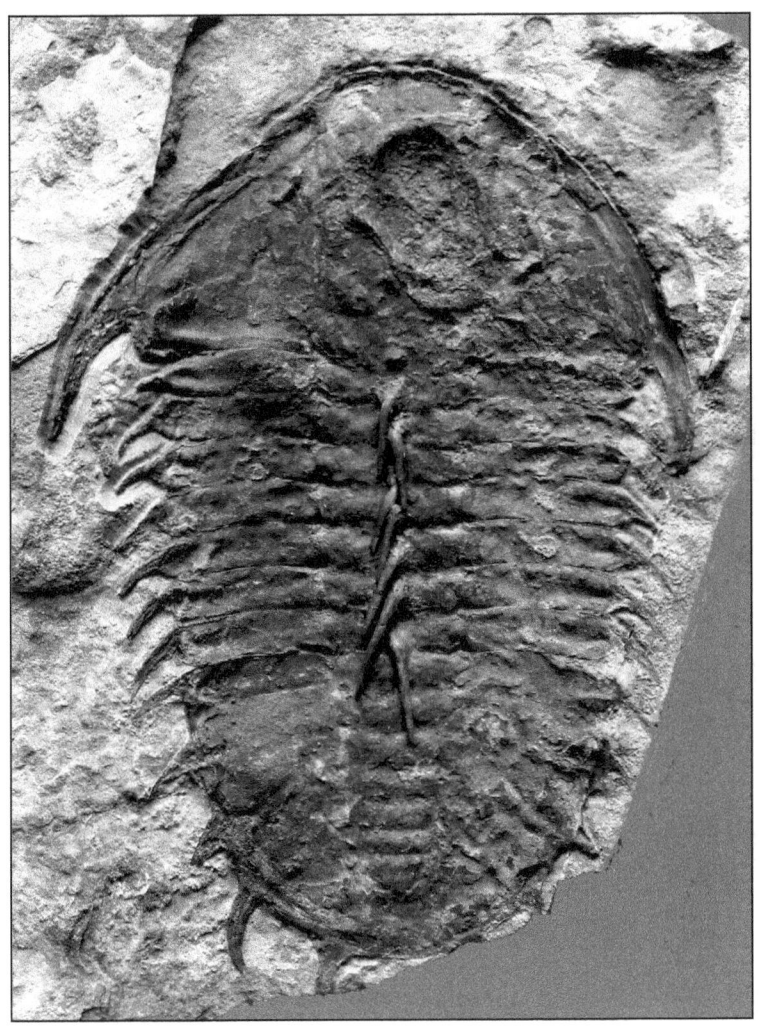

Olenoides superbus (Rasetti, 1946)
Marjum Formation
Length: up to 55 mm

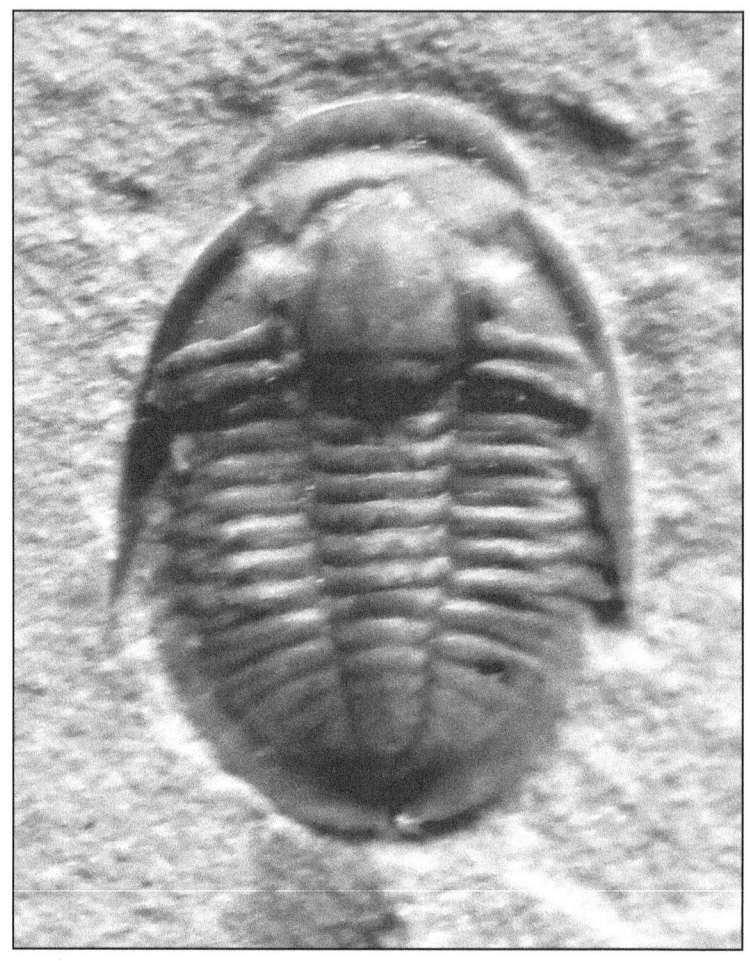

Cedaria minor (Walcott, 1924)
Weeks Formation
Length: about 14 mm.

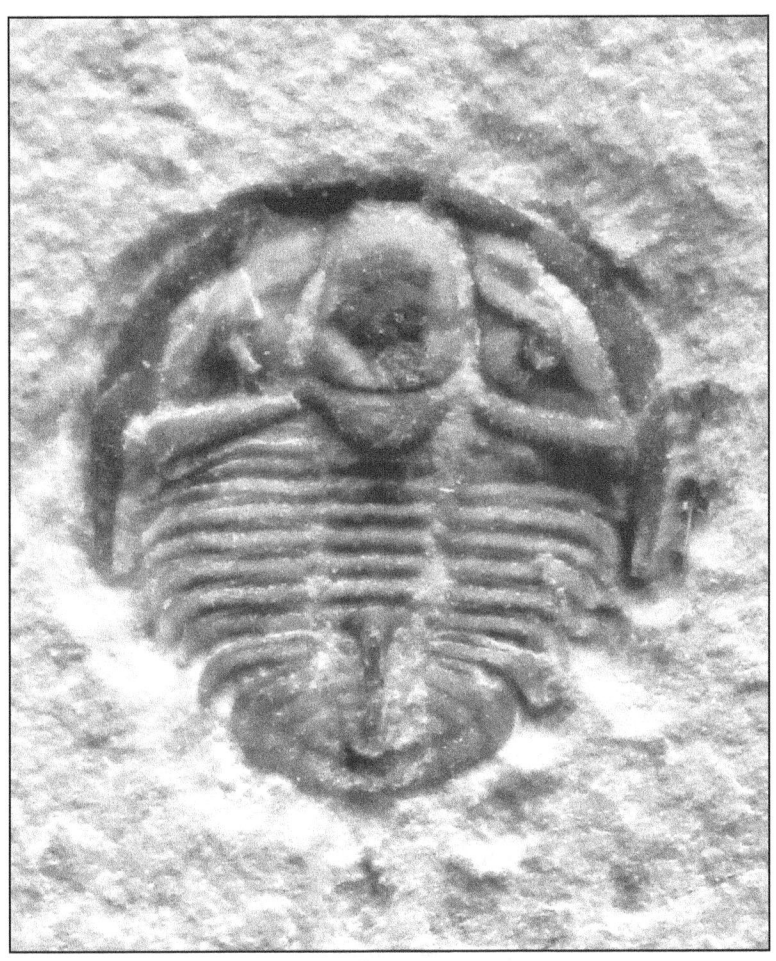

Weeksina unispinai (Walcott 1916)
Weeks Formation
Length: about 8 mm

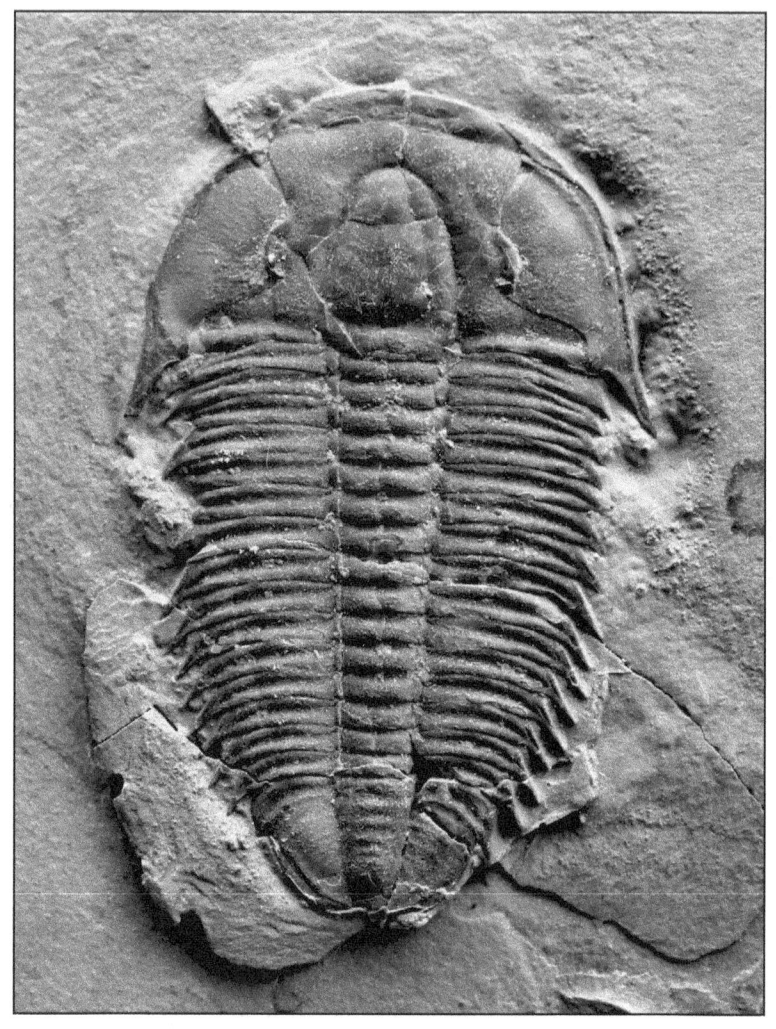

Utaspis marjumensis (Robison, 1964)
Marjum Formation
Length: up to 65 mm

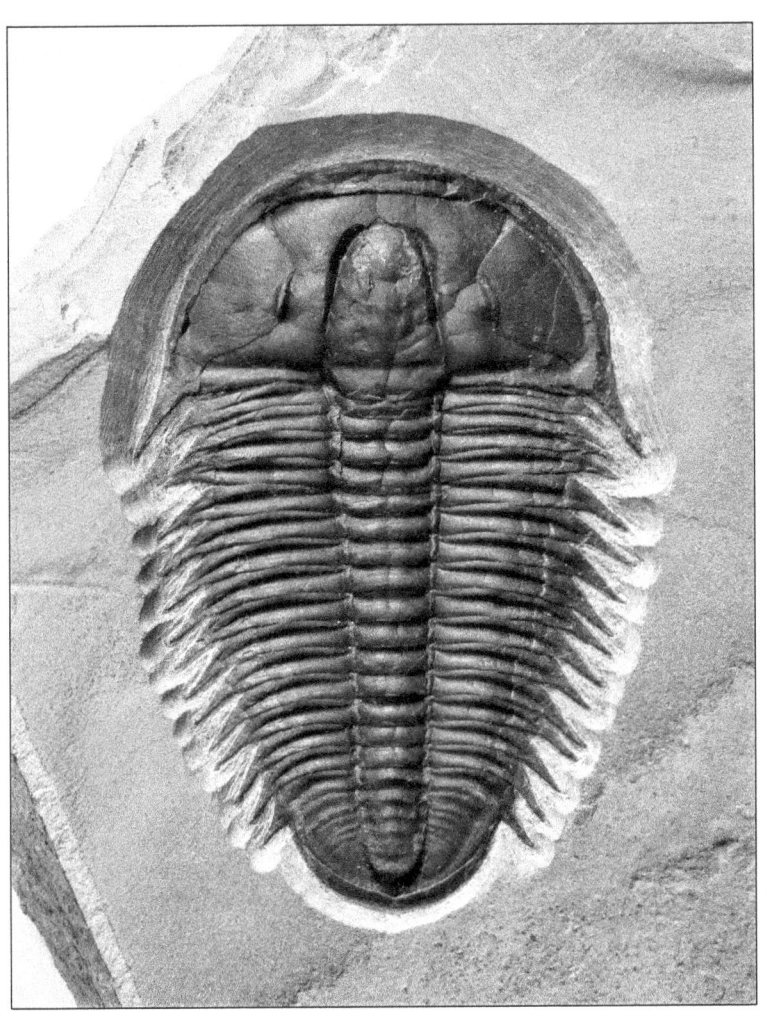

Utaspis marjumensis (Robison, 1964)
Marjum Formation
Length: up to 65 mm

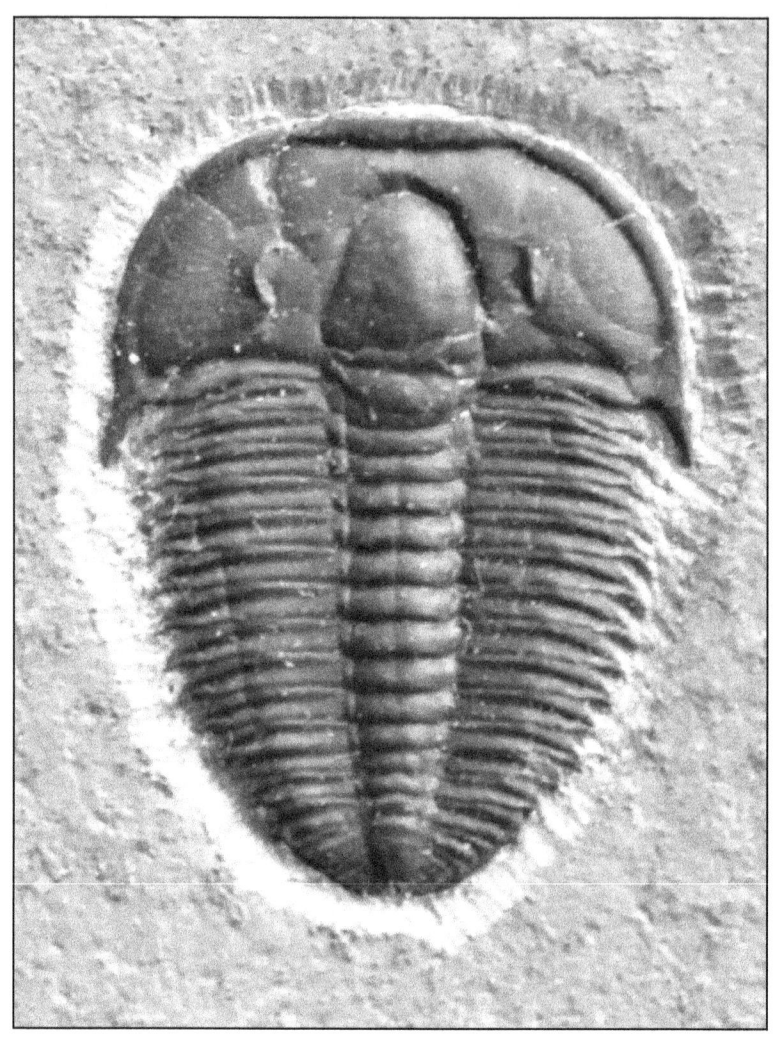

Elrathia sp
Marjum Formation
Length: up to 60 mm

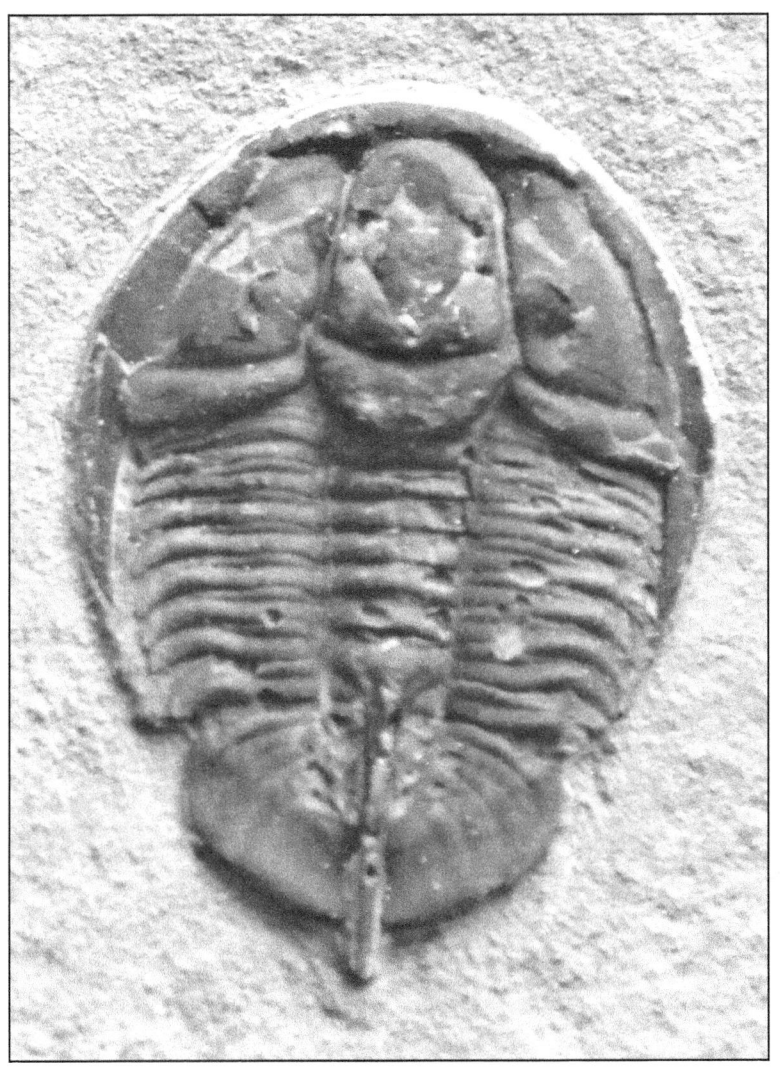

Genevievella granulata (Walcott, 1916)
Weeks Formation,
Length: about 28 mm

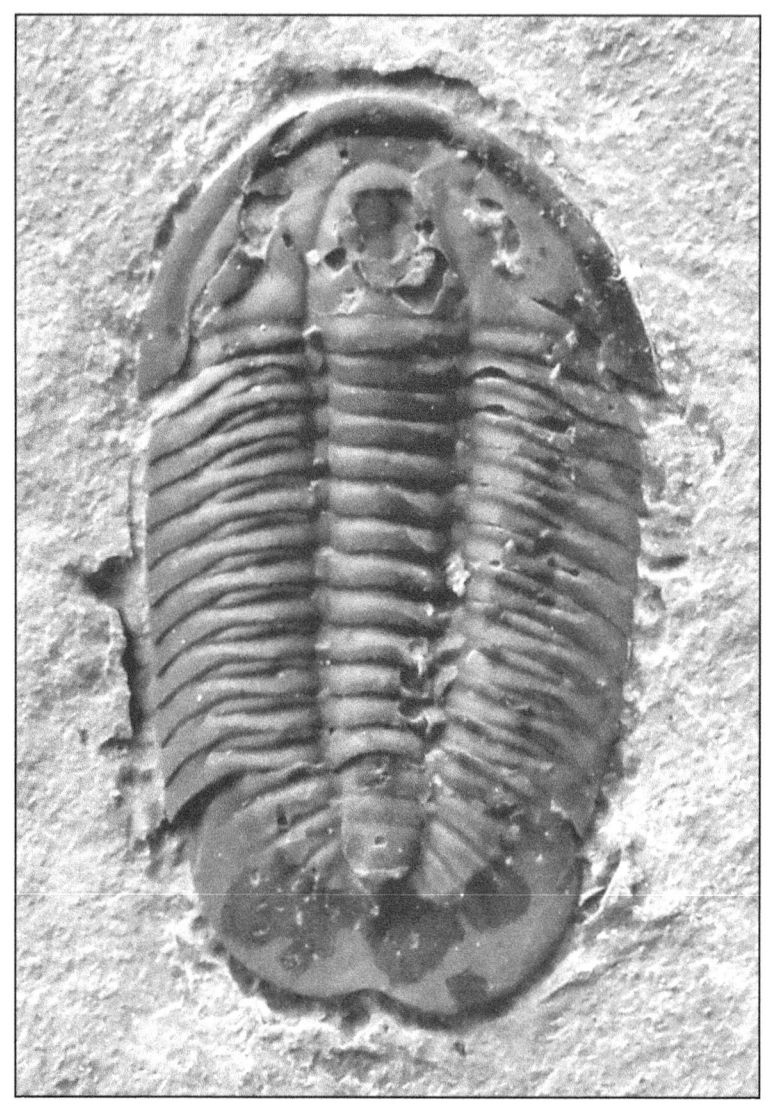

Coosella sp. (Lochman, 1936)
Weeks Formation,
Length: about 35 mm

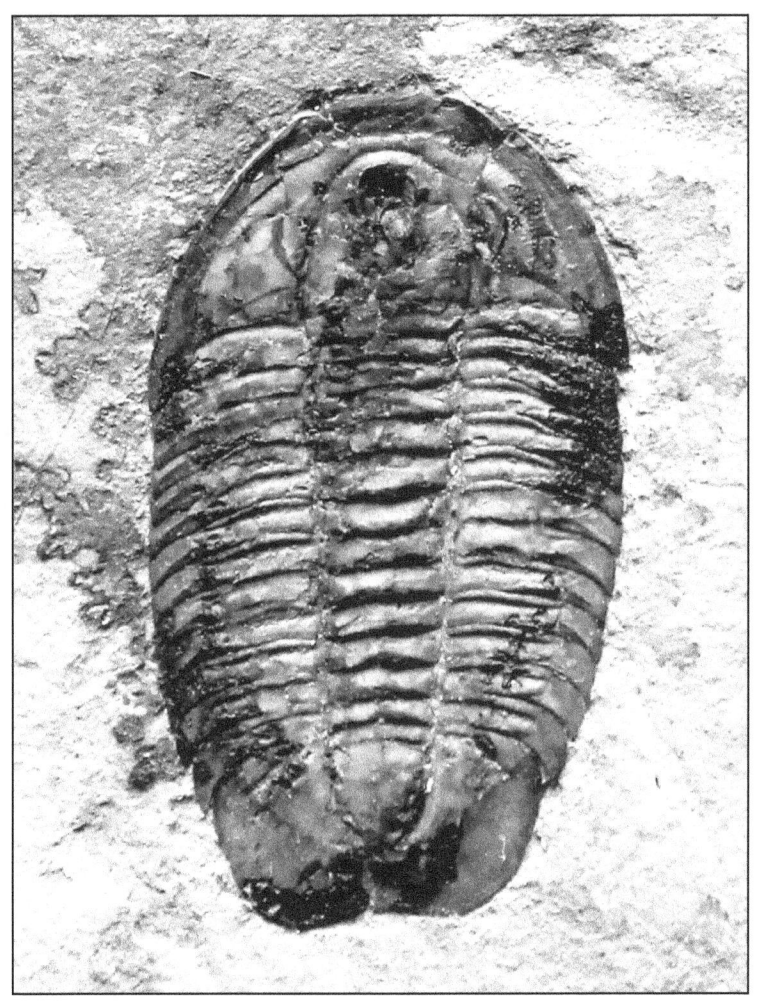

Coosella sp. (Lochman, 1936)
Weeks Formation,
Length: about 35 mm

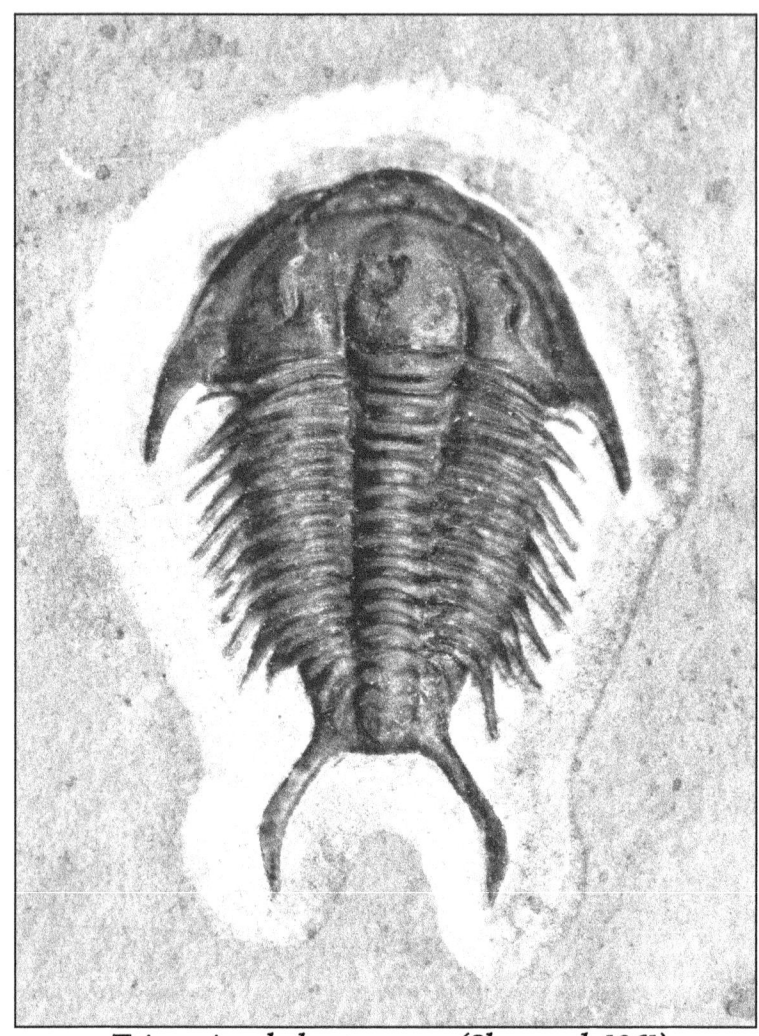

Tricrepicephalus texanus (Shumard, 1861)
Weeks Formation
Length: 40 mm

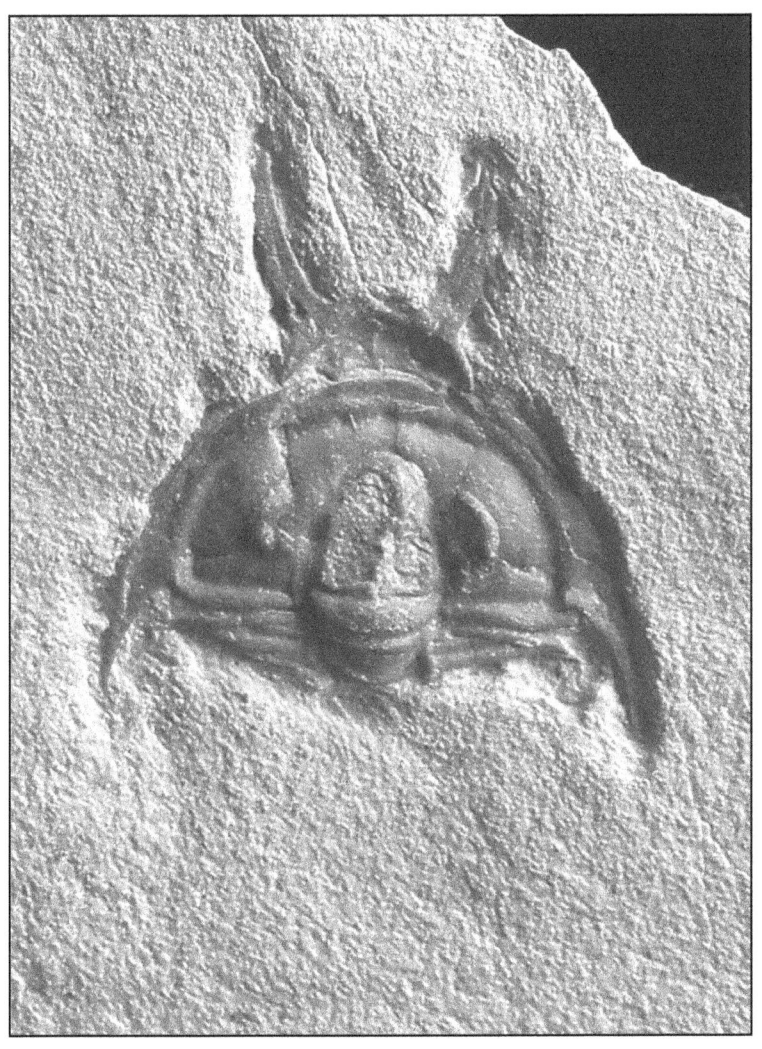

Tricrepicephalus texanus (Shumard, 1861)
Weeks Formation
Length: 40 mm

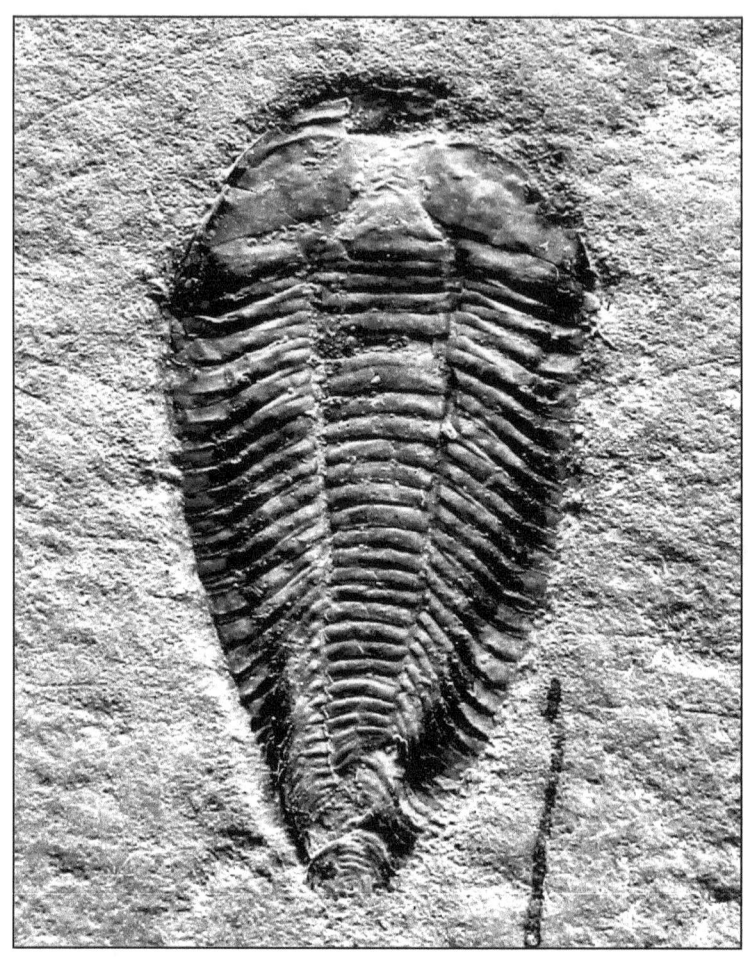

Nephalicephalus beebei (Peters, 2003)
Weeks Formation
Length: about 32 mm.

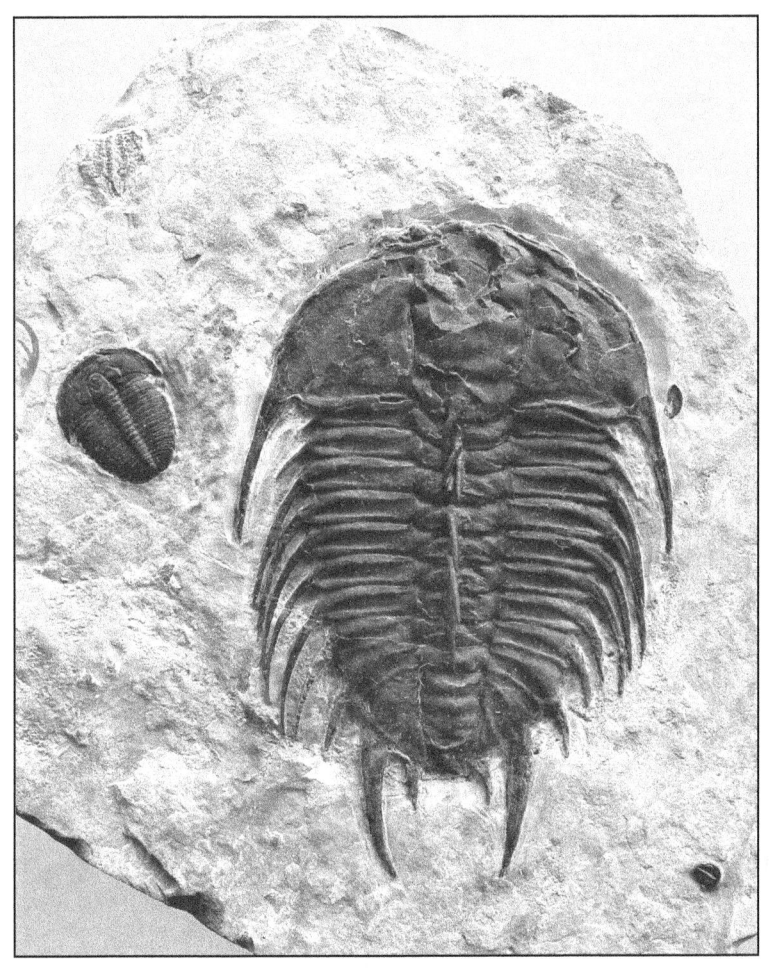

Olenoides nevadensis (Walcott, 1910)
Wheeler Formation
Length: 50 mm

Also, Elrathia kingi

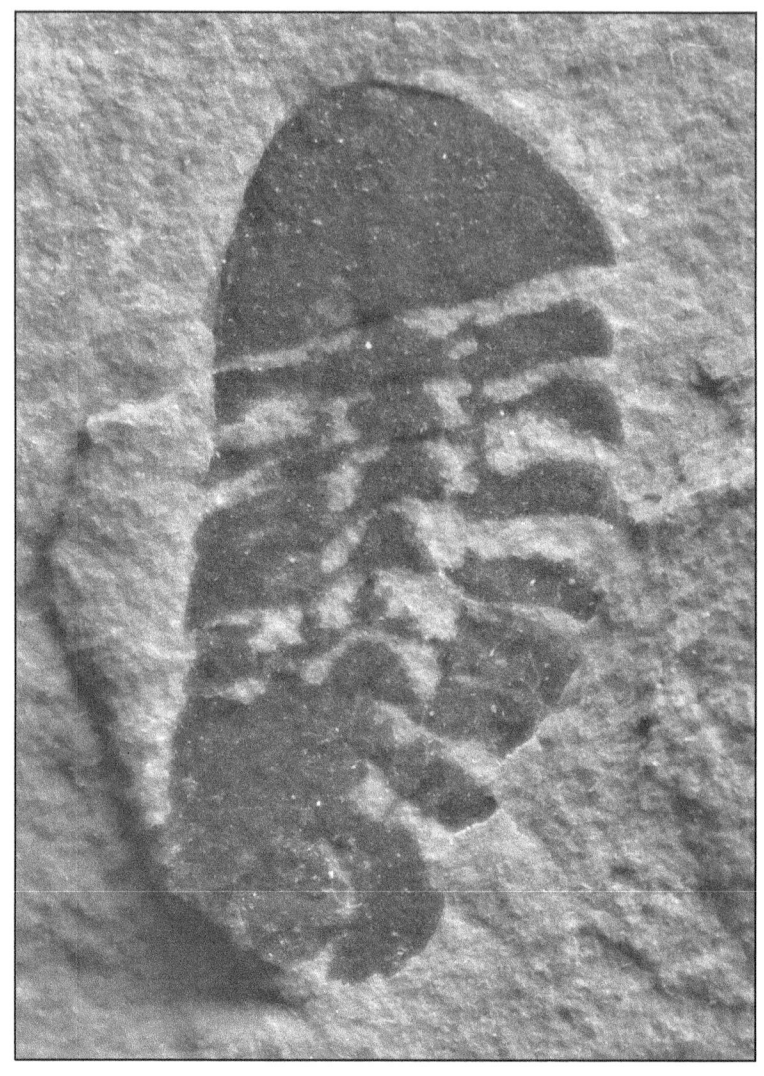

Aglaspida sp.
An Arthropod but not classified as a trilobite
Weeks Formation
Length: about 20 mm

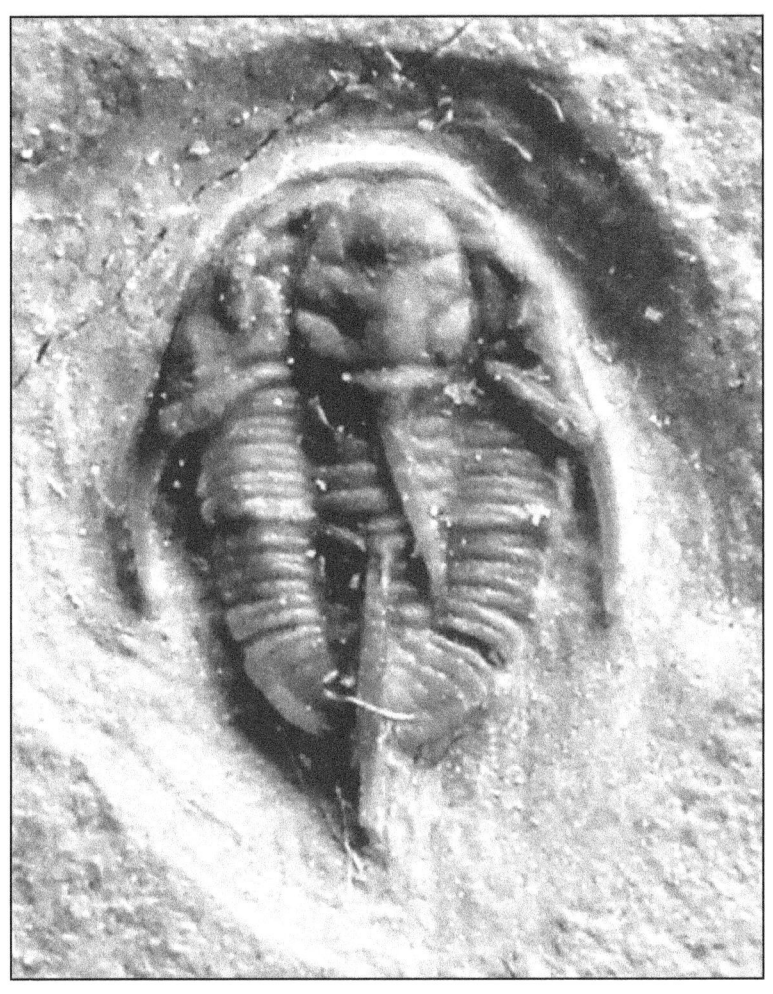

Lonchocephalus plena (Walcott, 1916)
Weeks Formation
Length: 6 mm

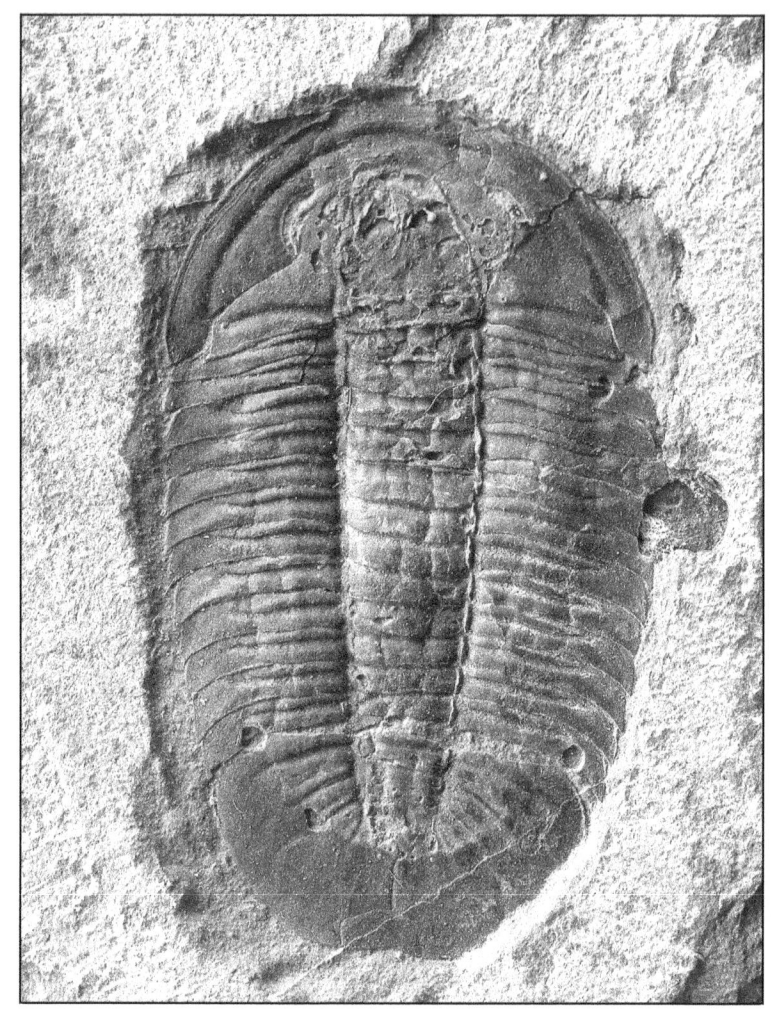

Syspacephalus sp. (Resser, 1936)
Weeks Formation
Length: 15 mm

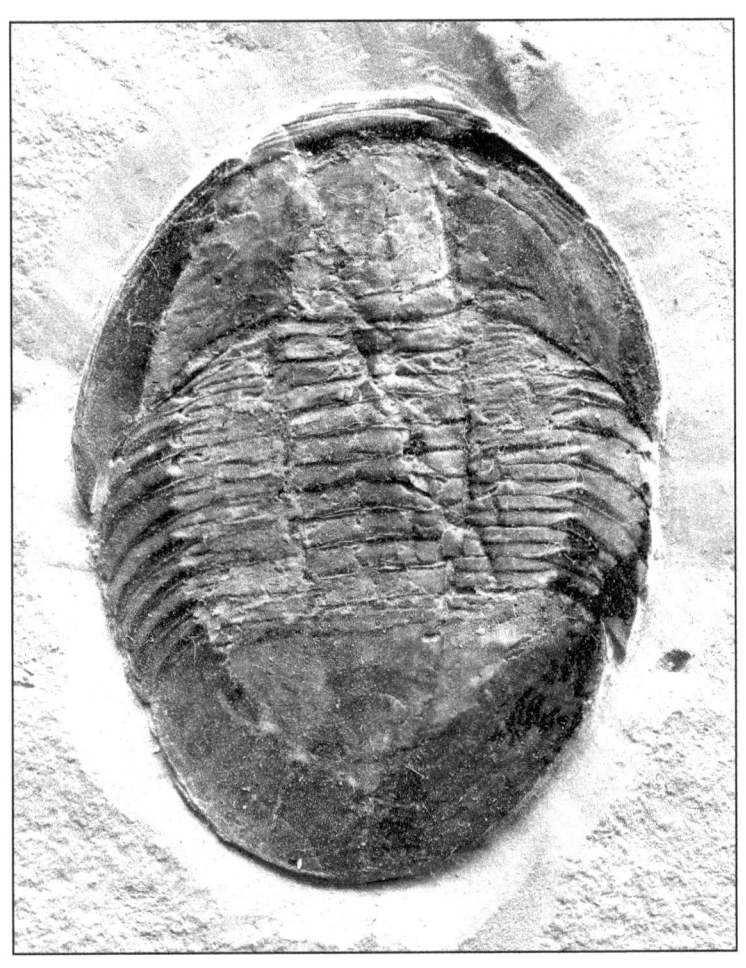

Selenocoryphe platyura (Beebe, 1990)
Weeks Formation
Length: 30 mm

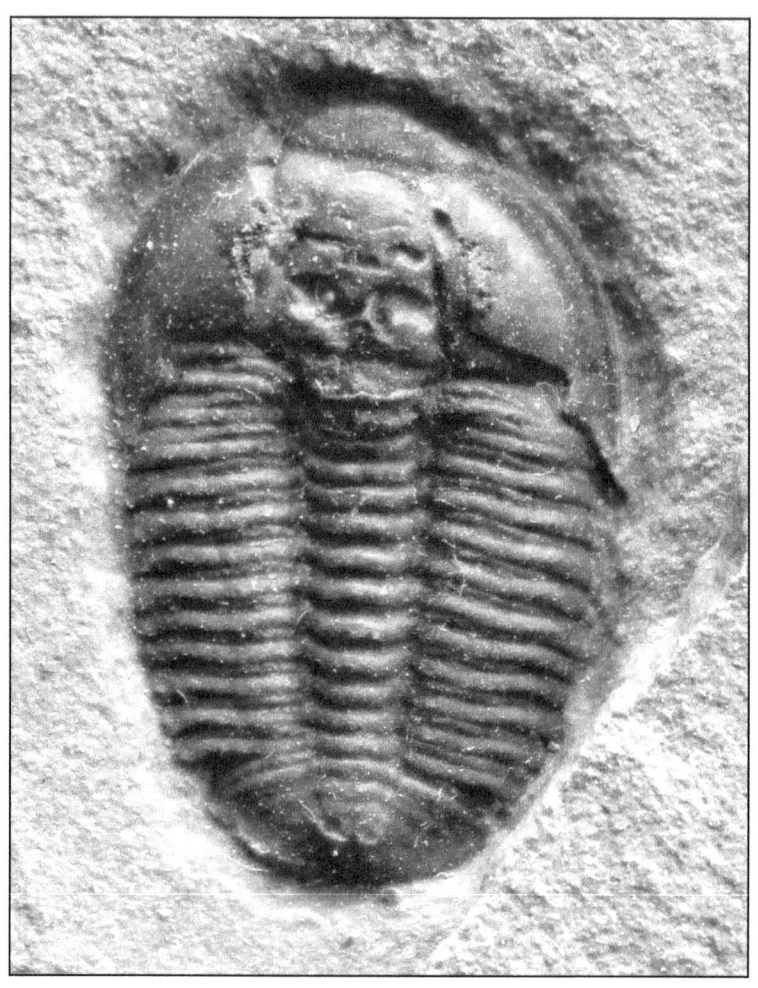

Modocia sp. (Walcott, 1924)
Marjum Formation
Length: 15 mm

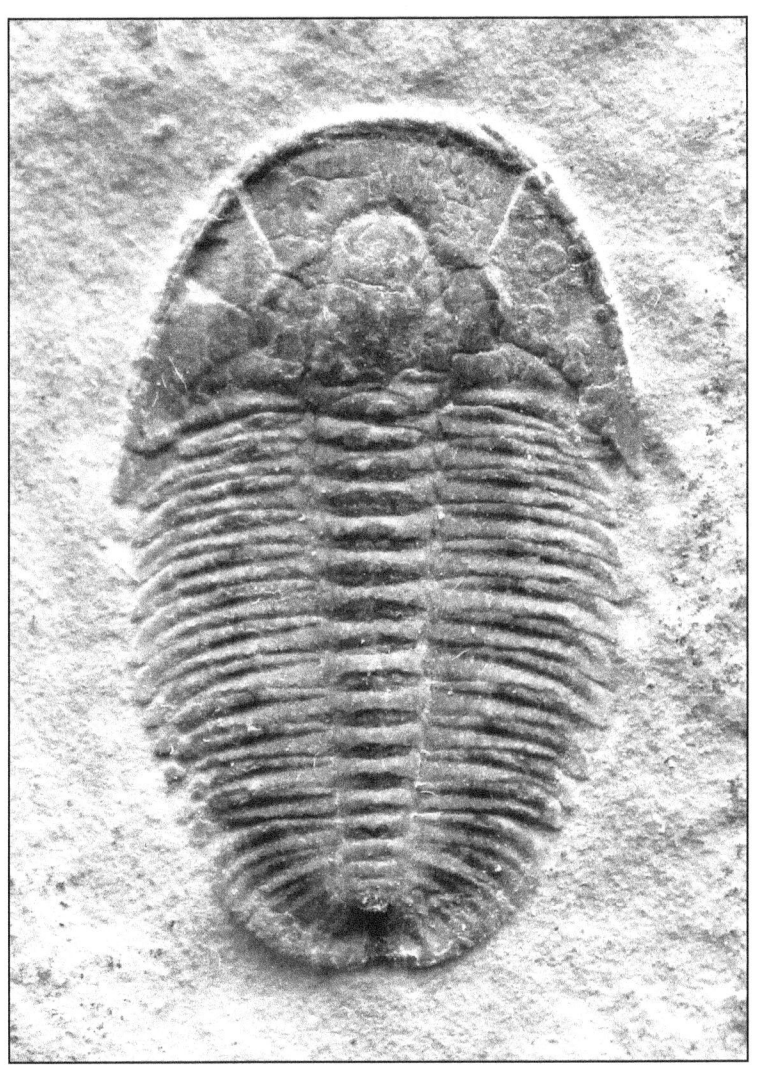

Modocia brevispina (Robison, 1964)
Marjum Formation
Length: 25 mm

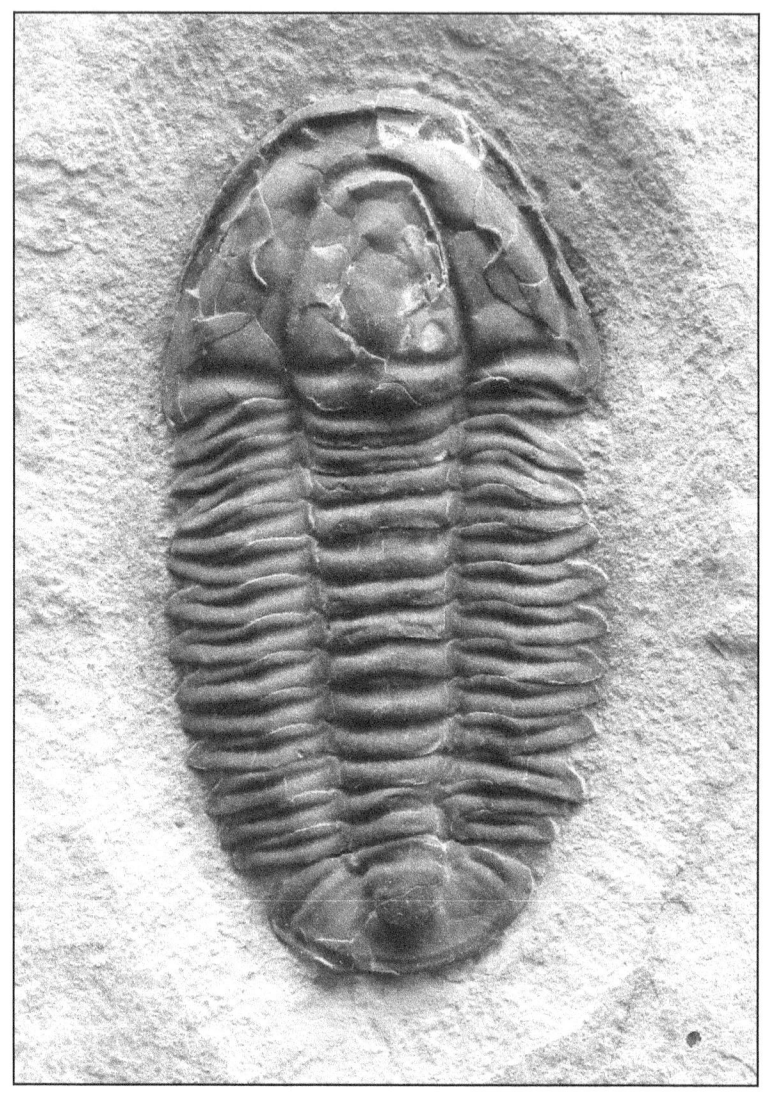

Modocia comforti (Robison and Babcock, 2011)
Weeks Formation
Length: 50 mm

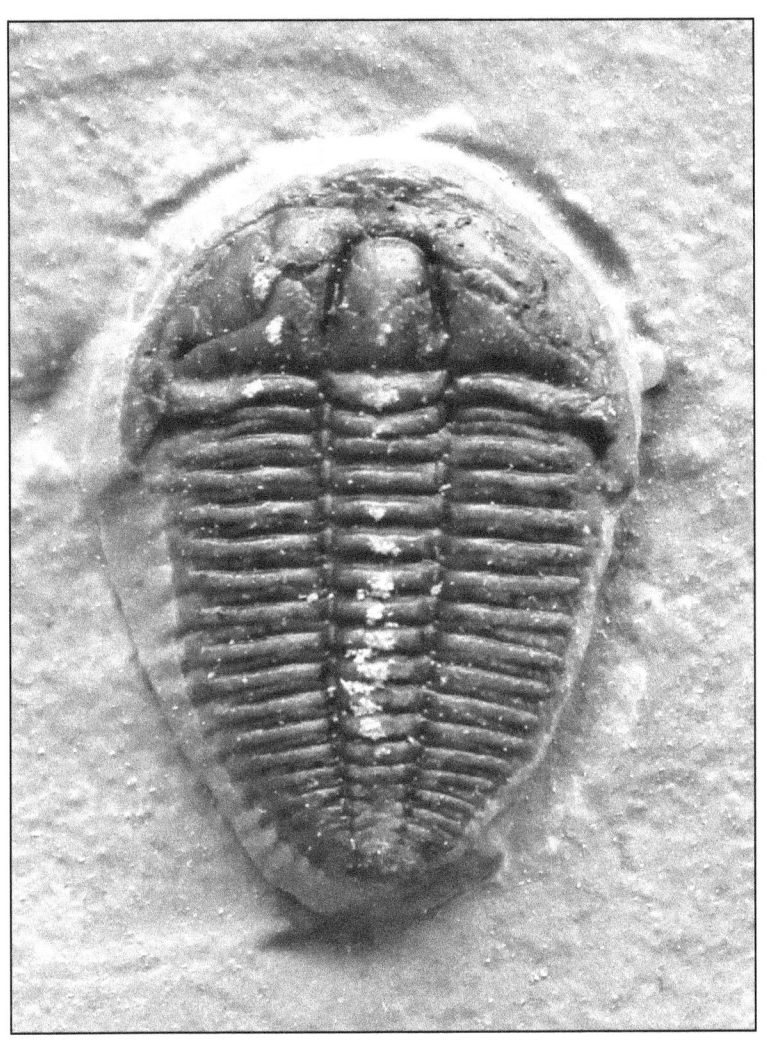

Bolaspidella sp. (Resser, 1937)
Marjum Formation
Length: 10 mm

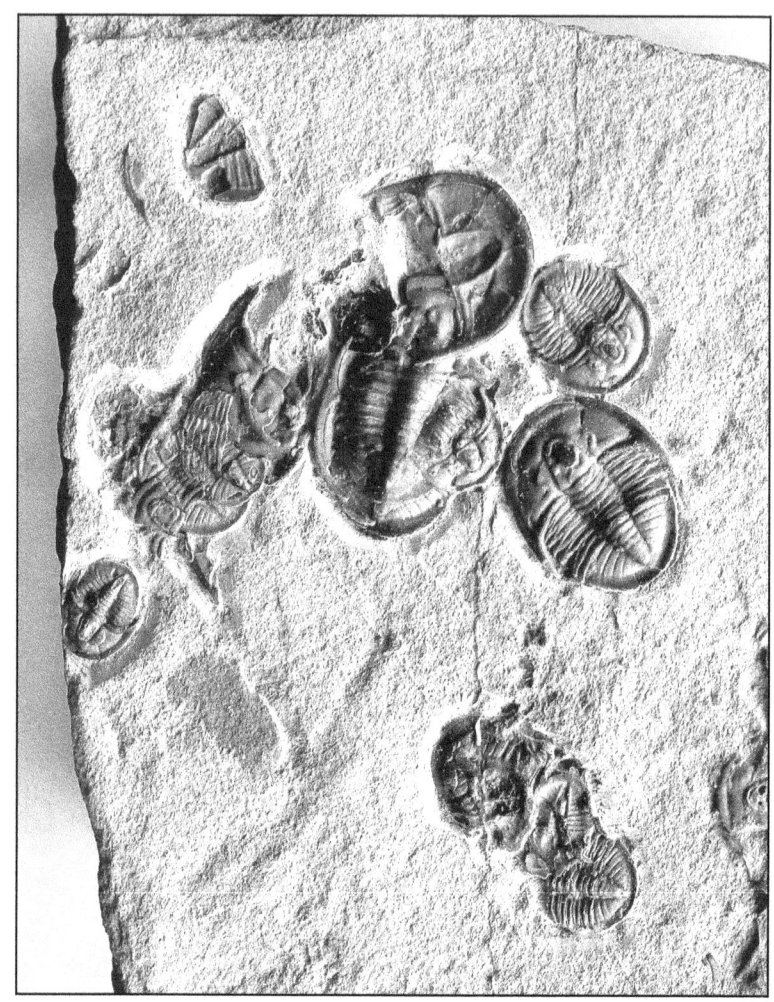

Cedaria minor (Walcott, 1924)
Weeks Formation
Length: 14 mm.

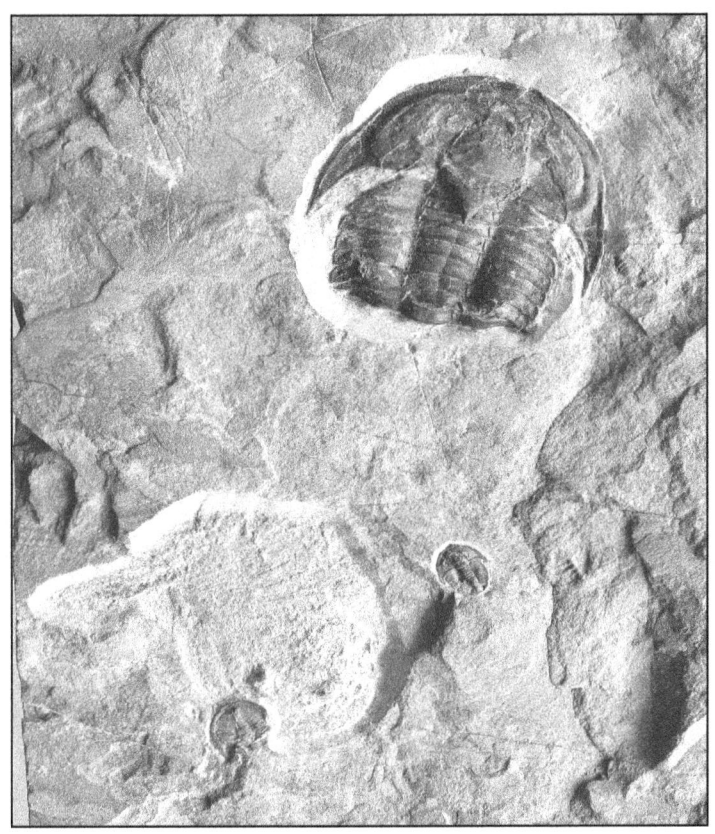

This Weeks Formation Plate Displays Four Specimens

Top: Arapahoia sp. (Miller, 1936)
Length: 20 mm

Center Right: Weeksina unispina (See page 51)

Center left: Beckwithia typa (Resser, 1931)
An Arthropod but not classified as a trilobite

Bottom: Cedaria minor (See page (68)
Weeks Formation

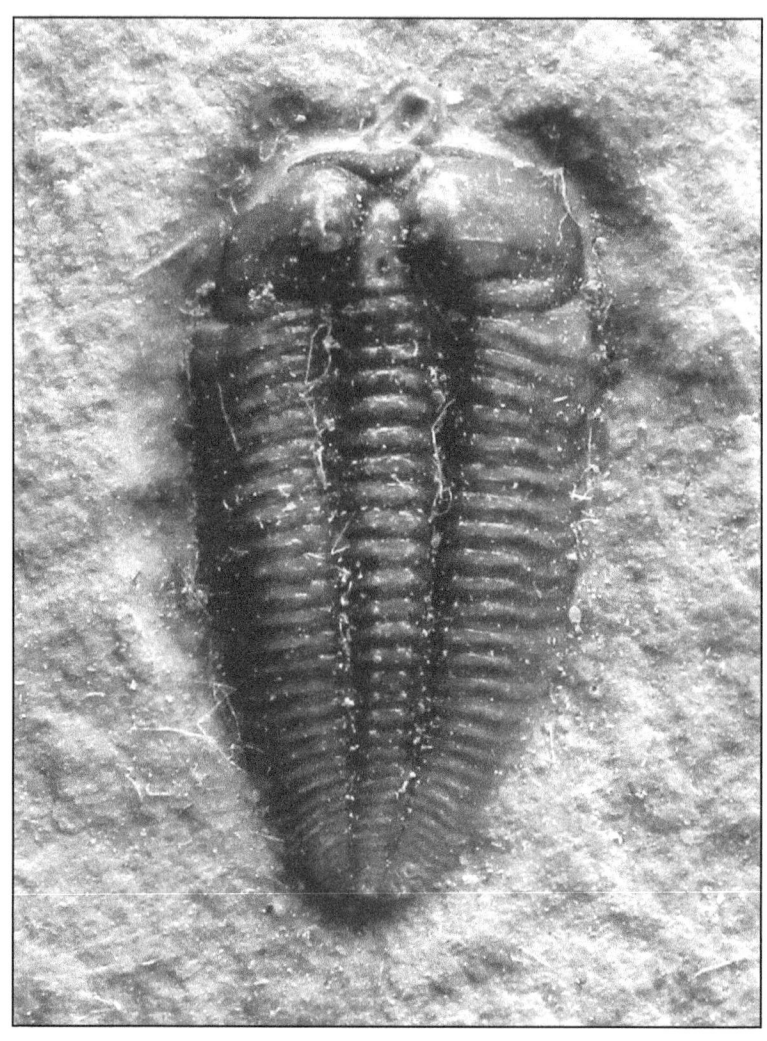

Dresbachia amata (Walcott, 1916)
Weeks Formation
Length: 10 mm

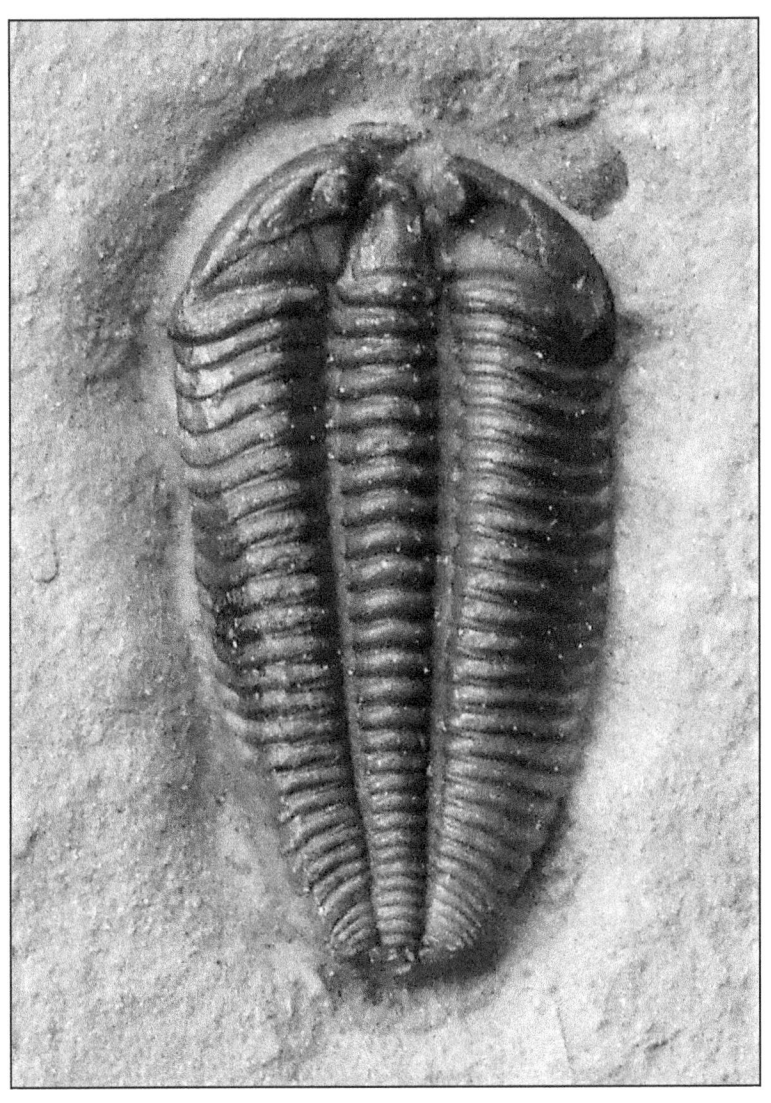

Dresbachia amata (Walcott, 1916)
Weeks Formation
Length: 15 mm

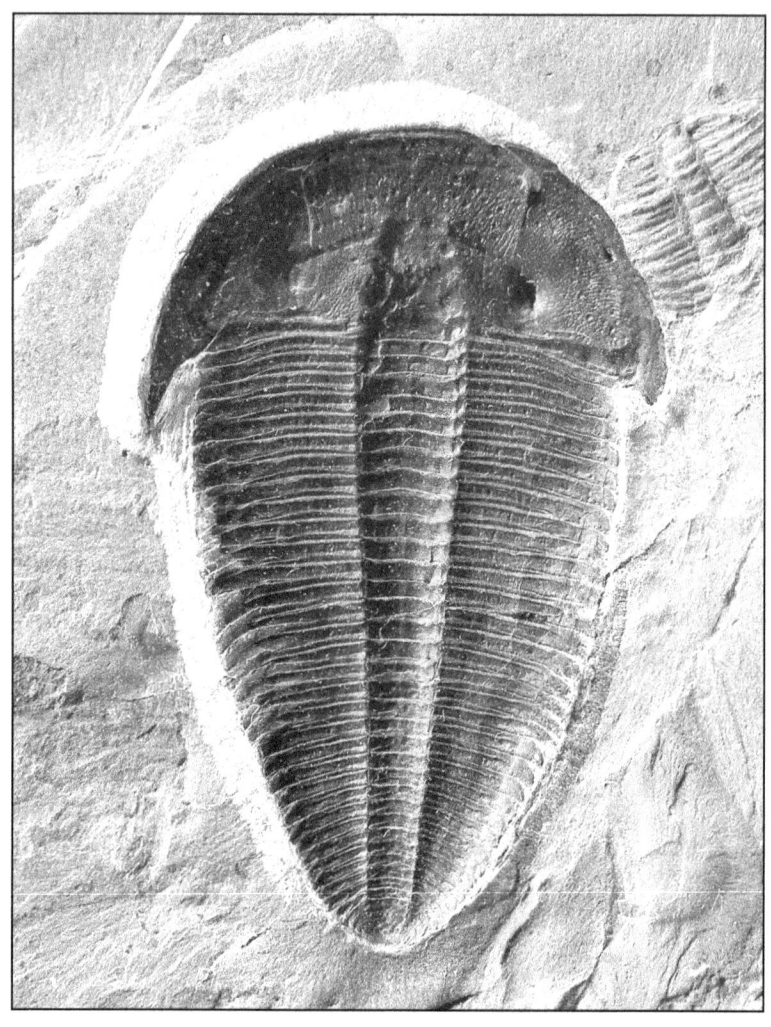

Altiocculus harrisi (Robison, 1971)
Wheeler Formation
Length: 50 mm
This specimen was unearthed by Robert Harris,
thus, the name "harrisi

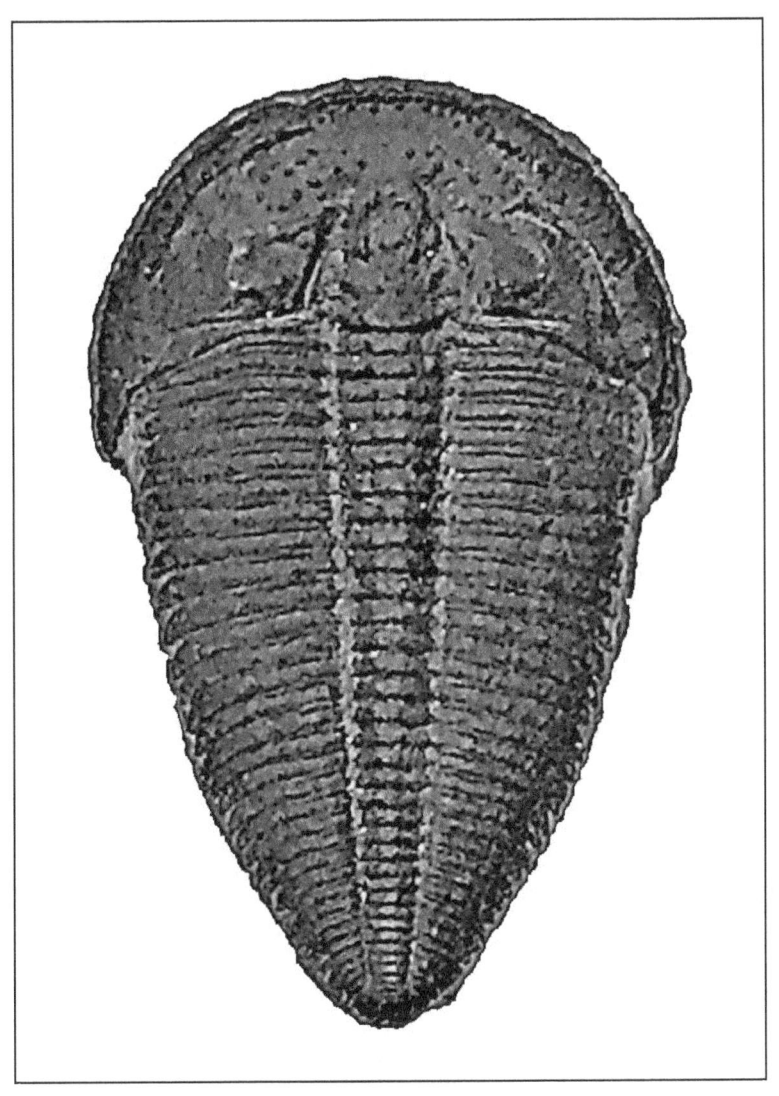

Altiocculus harrisi (Robison, 1971)
Wheeler Formation
Length: 35 mm

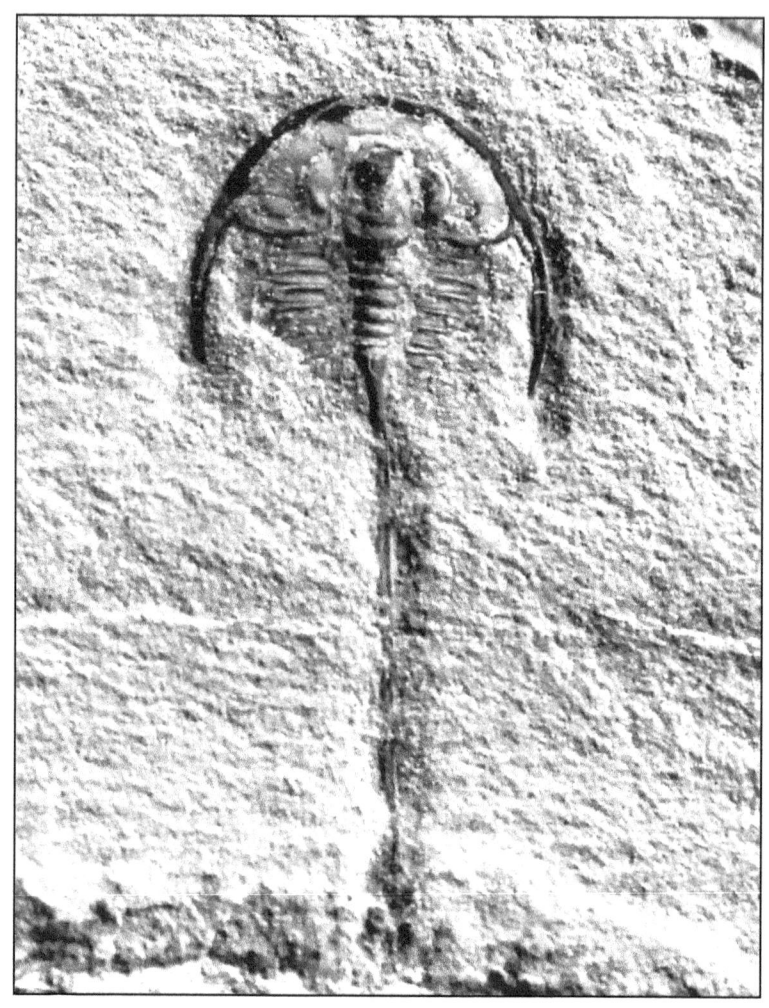

Gerospina schachti
Weeks Formation
Length: 37 mm

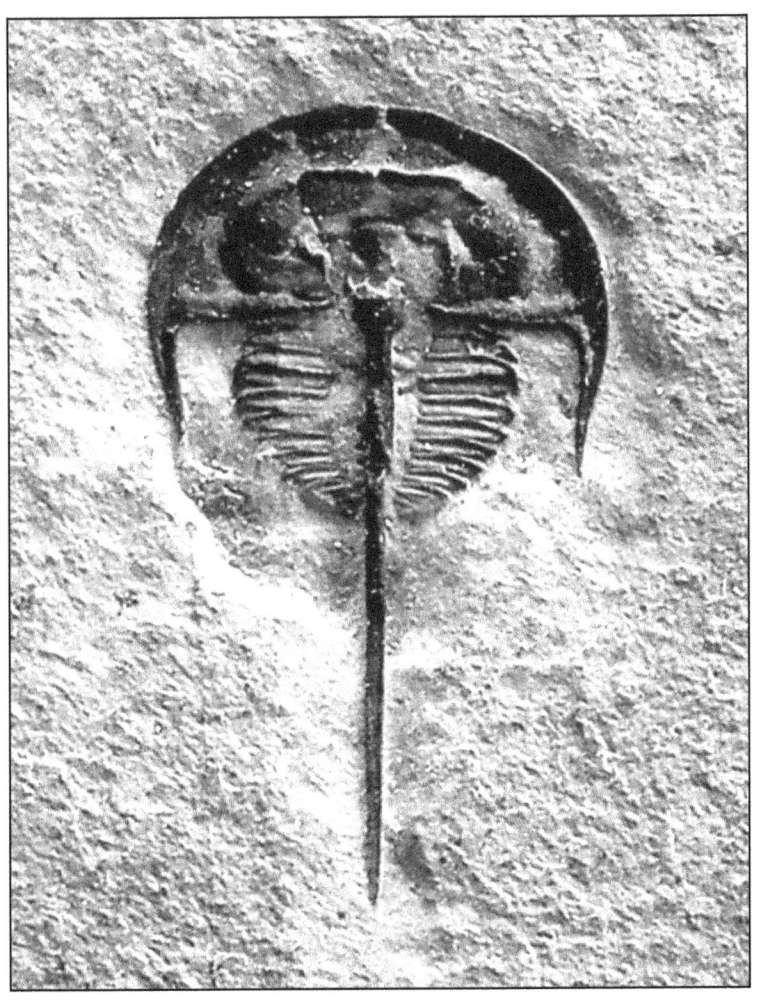

Norwoodia bellaspina (Walcott, 1916)
Weeks Formation,
Length: 18 mm

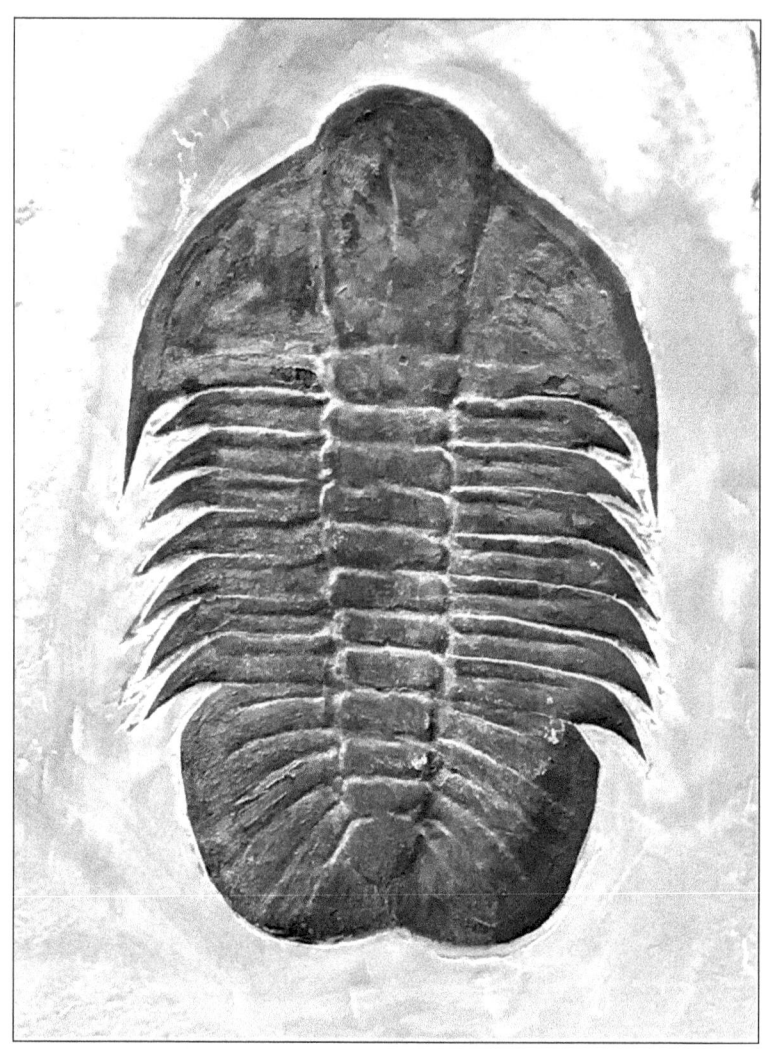

Hemirhodon amplipyge (Walcott, 1916)
Marjum Formation
Length: 85 mm

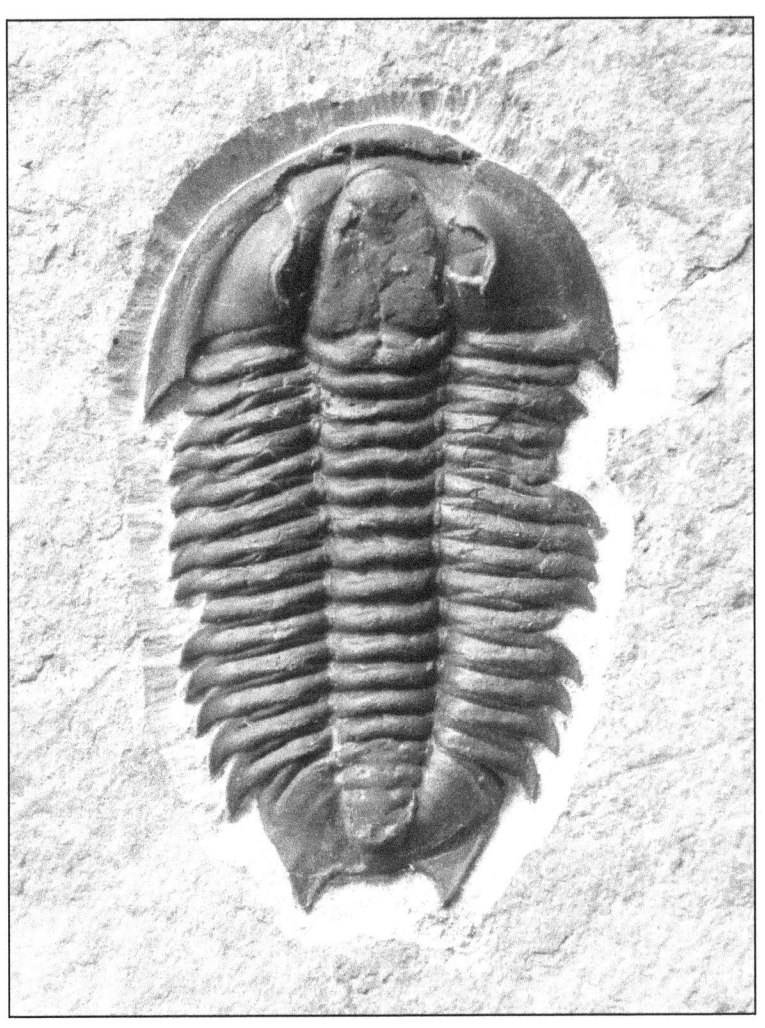

Meteoraspis sp. (Resser, 1935)
Weeks Formation
Length: 40 mm
One of the rarer Weeks trilobite species

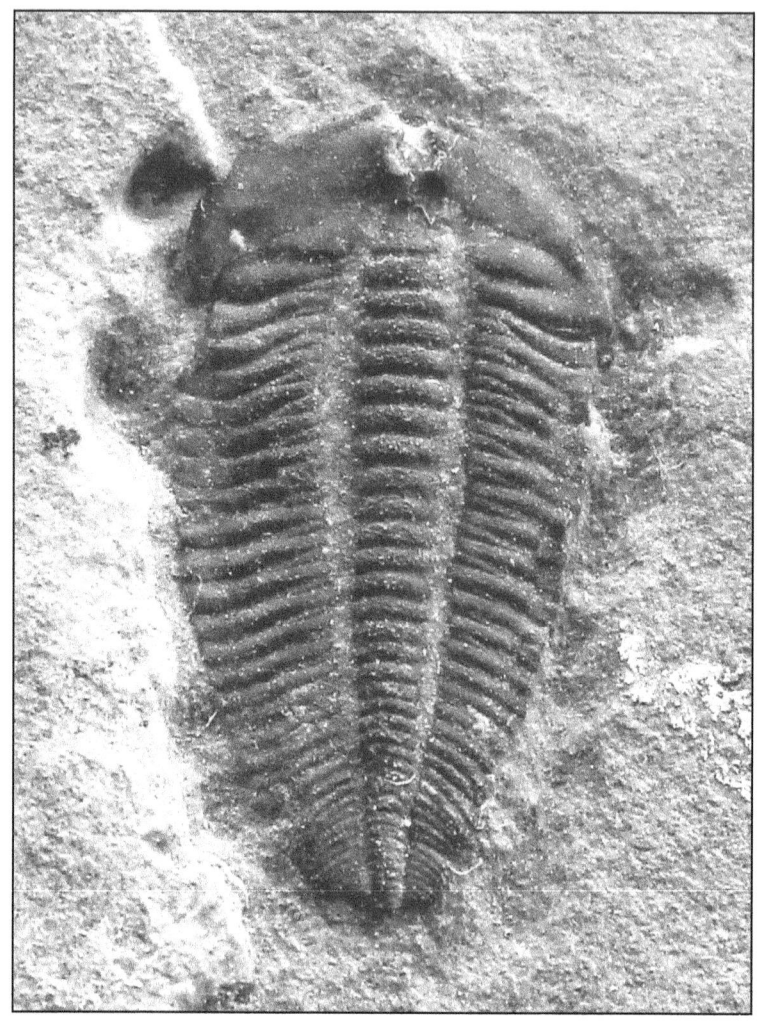

Menomonia semele (Walcott, 1916)
Weeks Formation
Length: 7 mm

Robert Harris Rock Shop History

For many years the Rock Shop operated by Robert and Iris Harris has been a fixture on Main Street in Delta, Utah.

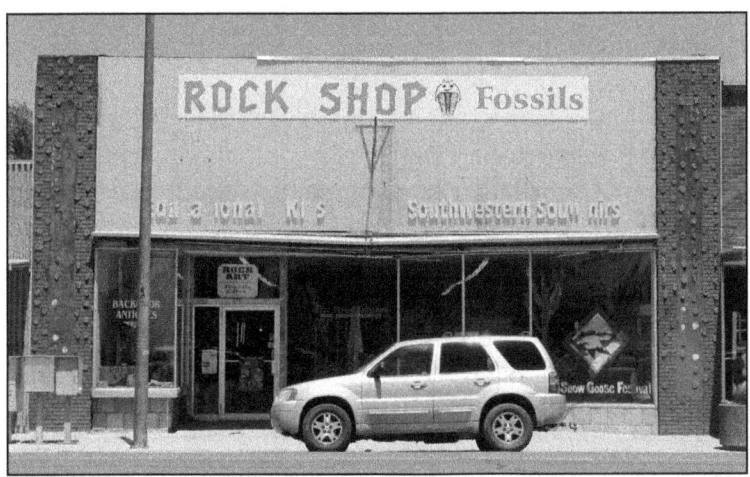

The beginnings of Robert's interest in fossils, gems and minerals was probably an extension of his father's interest in geology. As a youth, Robert and his siblings often accompanied their father as they explored the desert mountains west of Delta. Robert's father, Edward Daniel Harris, had worked in the Old Hickory Mine near Milford, Utah as a young man where he gained a knowledge geology and mining. Edward's favorite sites to visit near Delta included: Drum Mountain which had agate deposits, Topaz Mountain where topaz crystals could be collected and the Swasey Peak area of the House Mountain Range where several species of Cambrian-age trilobites were plentiful.

After his marriage to Iris Holman, Robert attended BYU in Provo, Utah for several years, returning to Delta during summertime where he worked at various jobs, one being collecting fossil trilobites from the mountains west of

Delta. Having acquired a fair number of trilobites he began selling the fossils to rock and mineral shops. Robert eventually dropped out of college and began mining fossils and minerals in a serious manner. Besides delivering fossil and mineral specimens to retailers, Robert started a mail order business. Because trilobites looked somewhat like a "bug", Robert named his business "The Bughouse".

For several years Robert partnered with his brother-in-law Loy Crapo. Eventually this partnership was dissolved with Loy's business retaining the Bughouse name, while Robert's business became West Desert Collectors. For some years Robert operated his business from his home. About 1970 Robert purchased the old DelMart Department Store on Delta main street where he has since operated his business.

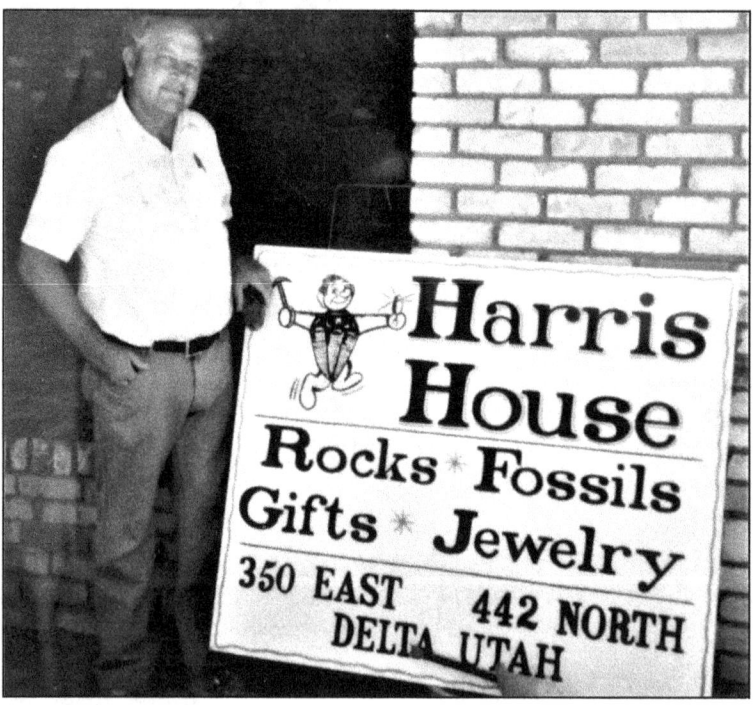

Throughout the years, Robert has travelled to most states in the country calling on owners of rock and mineral businesses and currently has contacts with hundreds of dealers in similar businesses where he sells and also trades for fossil and mineral specimens that are not found in Utah.

Some of the fossil and mineral specimens that are collected in and around Utah are: several species of trilobites that are found in the Wheeler Shale and Marjum Limestone rock formations in the House Range west of Delta; topaz crystals from Topaz Mountain northwest of Delta; snowflake obsidian south of Delta; red beryl crystals in the Wah Wah Mountains south of Milford, Utah; geodes from a site near the Dugway Proving Grounds in Tooele County north of Delta; various species of fossil fish from several sites in the southwestern part of Wyoming in the Green River Rock Formation; septarian nodules (geode-like) from a site near Glendale, Utah and also Picasso Rock, which Robert named, from Beaver County, Utah.

Robert and Iris often travel to various parts of the country to sell their ware at gem and mineral shows. One market is in Quartzsite, Arizona where there are several markets held during the winter months. They also usually spend about a week in Tucson, Arizona during the month of February where the world's gem and mineral and craft dealers buy, sell and trade a wide array of goods.

Visitors to Robert and Iris's Rock Shop in Delta will not only find rock and mineral specimens from Utah and nearby, but specimens from around the world. Browsing the display cases, shelves and tables filled with assorted natural "earth" materials is like being in a geological museum.

Excerpts from Some of Robert's Writings

1960-1961: Found fossil trilobites being sold in a southern Utah rock shop. They wanted to buy more! I rushed home and asked my father to show me where trilobites came from. That began a seven-year companionship with my dad. I worked for one week installing heating ducts in the new Delta High School, but it was colder in that building than it was out at the trilobite mountain, so I returned to collecting.

In the fall of 1960, I became interested in collecting and selling trilobites. My father accompanied me to the desert and after a while I learned how and where to collect them. It seems like we found sale for them right away.

In the summer of 1961, I took my first trilobite selling trip. I went to California for six days. I sold $700.00 worth of trilobites. Expenses in those days were low: gasoline was 25 cents per gallon, the car went 28 miles per gallon, so the vehicle expense for the trip was about $15.00. Motels were about $8.00 per night for a total of $40.00. All expenses were near $65.00 for the week, leaving a profit of about $600.00 for three weeks of collecting and selling. That compared with about $60.00 per week I could earn working for wages. I was hooked!

My Unique Successes as a Rock Dealer
Robert L Harris

In my years as a rock dealer I've experienced three singular successes. **First**: In the 1970s the metaphysical movement gained wide popularity. It began by certain authors writing about claimed success of using quartz crystals and other minerals in healing all sorts of maladies, both physical and mental.

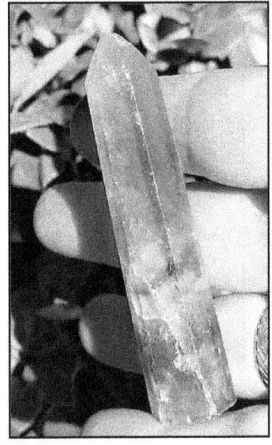

The excitement and interest was rather intense, especially regarding the healing qualities of quartz crystals. Lots of money was exchanging hands. To participate in the frenzy, I began cutting other materials into the shape of quartz crystals; fluorite, turquoise, rhodochrosite, and others, pieces to be held in the hand and used as a massage wand and later smaller pieces to be crafted in pendants.

Our success was immediate. For a short time I had four men cutting and polishing wands. The wand business lasted less than four years, when cutters from Brazil and China flooded the market with large quantities of similar and widely greater variety of like items at about one-fourth the price I was charging.

Second: I was in a quandary as to know what to do with the cutters in my employ. But shortly the problem was solved. About 1980, while making a delivery at Fisher's Rock Shop in Orderville, Utah, a native American woman came into the shop to offer for sale her wares. One of the items she was selling was a wooden carving of a bear having very few details, basically just a silhouette. She gave me permission to trace the outline of it. I returned to Delta and with one of my cutters, Felipe Barrera, cut a stone bear of Picasso marble. Again, there was immediate success. I kept all the cutters working. Later we began cutting bison silhouettes. Today we still sell bears and bison, only on a much smaller scale.

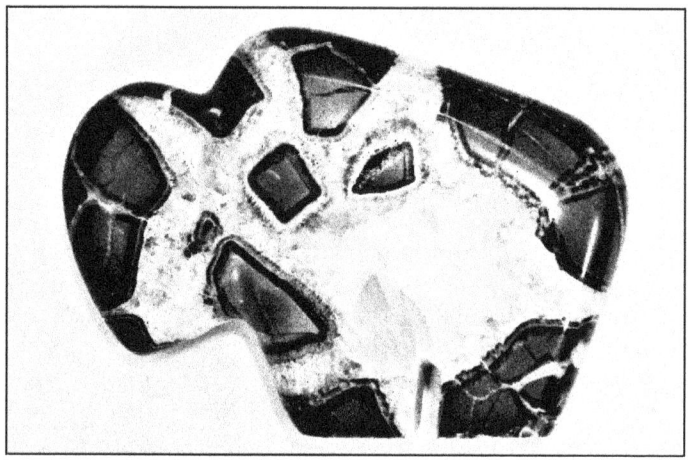

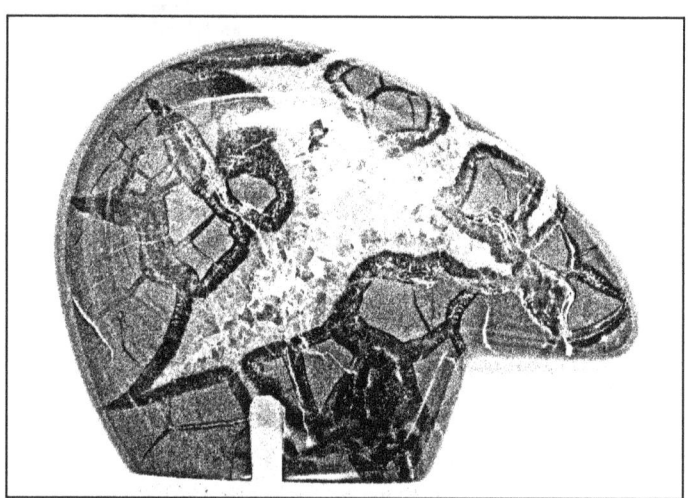

Mark Harris Photos

Mr. Barrera and others established large cutting shops in Mexico and have the bulk of the business. These two shapes (bear and bison) are cut all over the world; China, Mexico, Pakistan and other countries.

Third: Another successful venture is making carvings such as spheres and eggs as well as hearts and animal shapes around the crystal cavities in Utah septarian nodules. This work is also copied around the world with geodes and other crystal rocks. All these items above are found in a very large percentage of the rock and mineral shops around the world.

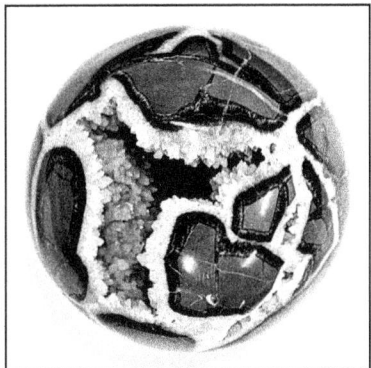 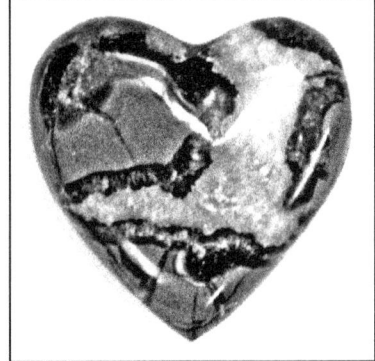

Extra Success

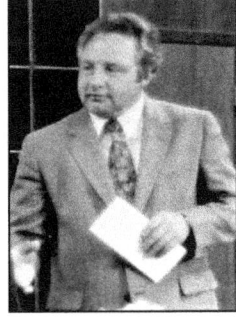

I served ten years on the Millard County School Board, 1970 to 1980. I was president of the board when the middle school was built in Delta. I was elected as president of Utah School Board Association but lost the local election in 1980 and therefore never served on the state level.

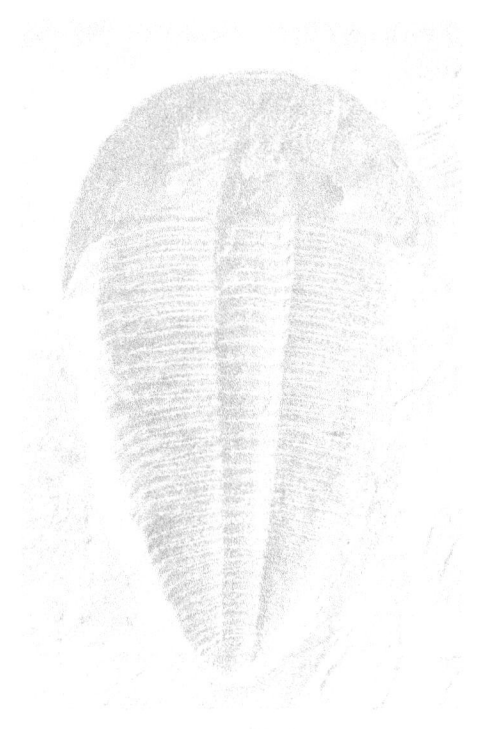

Reviews from West Desert Collectors Website

A very interesting shop for rock and fossil lovers like me. The manager was very informative and told us where to go to do our own digging/rock hounding. If I fail to find fossils or desirable rocks I will return to buy here.

"Fascinating and wonderful place to explore!"

I spent about 2 hours here with a teenage boy who was just as enthralled as I was by the variety, sizes, and great prices of millions of rocks! Don't plan on zooming in and out of here-you'll be spellbound by all there is to see! WONDERFUL place!

"Wide Selection of Rocks & Fossils"

This rock shop is easily located on Route 6 on the West end of town. I stopped in on a whim and am glad I did. The store has a diverse variety and large inventory of local and worldwide fossils and stones. Many are polished and some are in their natural state.

The owner is both friendly and knowledgeable. He explains not only what the stones are, but their local origins, where else they can be found, and what makes them special. Prices are reasonable.

I bought several eye-catching and unique items. A great spot for both collectors and casual customers.

"Best Rock Shop in Utah"

West Desert Collectors is the best rock shop in Utah in my opinion. Bob has a great inventory for fair prices and is willing to haggle (a little bit) with fairness. Ask him about any rock/mineral and he has a vast knowledge of them and if he has them in stock. Ask Bob or Jay good sources for rocks/minerals in the area, and they will proudly point them out. We'll be back.

"Fun rock shop"

After (unsuccessfully) digging for topaz in the area, we stopped at this rock shop to check out their goods. They had a wide range of fossils and rocks for sale. The owner, Bob, is a nice guy willing to let you browse for hours or help you find an unusual specimen. We look forward to seeing them again at the gem shows.

This is the best rock shop in Utah. The owners are very helpful, and willing to tell you about everything. I look forward to stopping here whenever I am in Delta.

A diamond in the rough in this small town. Wow-great specimens at fair prices! The owners are the real deal when it comes to being educated about fossils and rocks. Also, great jewelry!

Best Rock Shop in the West!

Visited West Desert Collectors rock shop while seeking trilobites. WOW! What a fantastic experience! The absolute, bar none, best customer service we've received anywhere. Small town America treats you right and this business is a pure example of the best of the best. Bob Harris and Jay Spor showed us the best time we could have had in the mecca of Rock and Gem hunting-Delta, Utah! Don't pass through Delta, UT on highway 6/50 without visiting, you will be pleased you stopped!

They have one of the larger selections of rocks I have seen at a rock store (and I've been to rock stores in several different states).

Really nice people. I probably was in their store for over an hour looking through boxes and boxes of rocks and they answered all my questions and helped me find lots of rare finds.

I left with some rocks that I had not seen before...and a gorgeous piece of Tiffany stone--I will definitely be coming back here when I am back up that way...

If you need it, they have it...rocks, fossils, jewelry, t-shirts, trilobites, books and everything imaginable.

Bob is the greatest guy! He doesn't "computer", but if you go to his store you will be enriched with information beyond what you expect! We've been visiting him for 10 years and our rock hounding has been more "finding" than looking because of his willingness to share! Besides, he has it all in his shop, reasonable prices and better specimens than I have found on my own!

Great shop and a great guy! Bob knows the west desert like no others. Just a pleasure all of the time!

Awesome place to check out and buy some minerals!

Great place! Bob was a pleasure talking too and even signed my shirt!

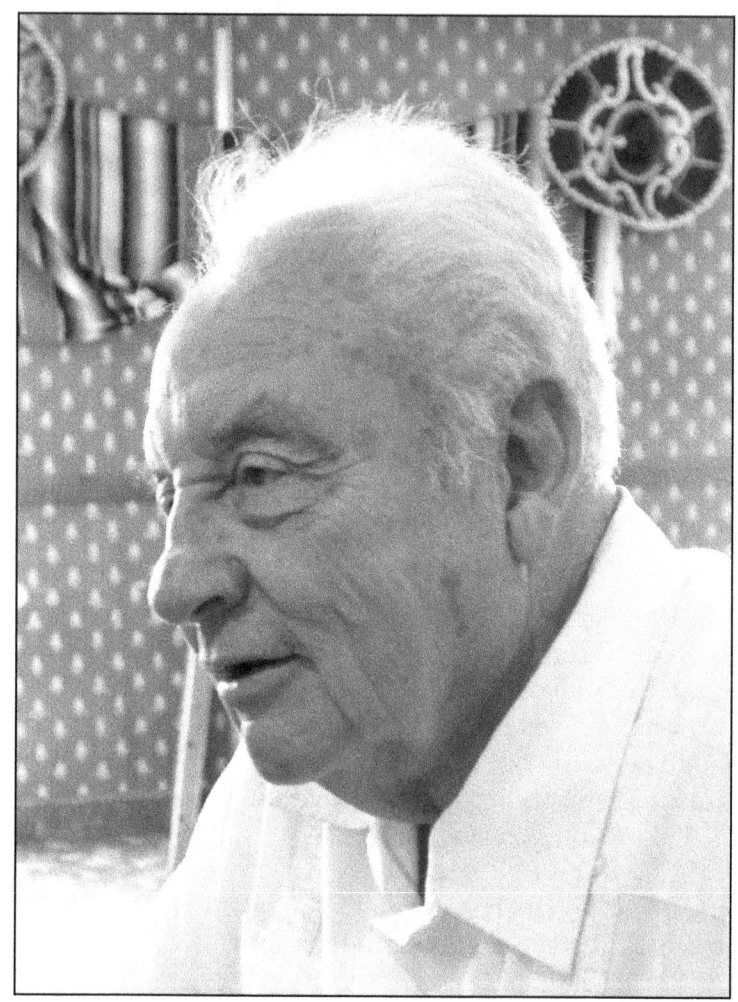

Robert L. Harris

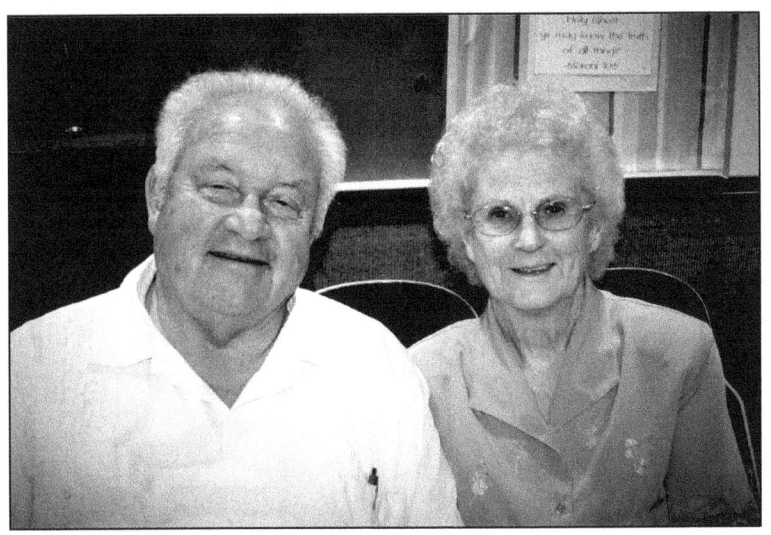

Robert and Iris Harris

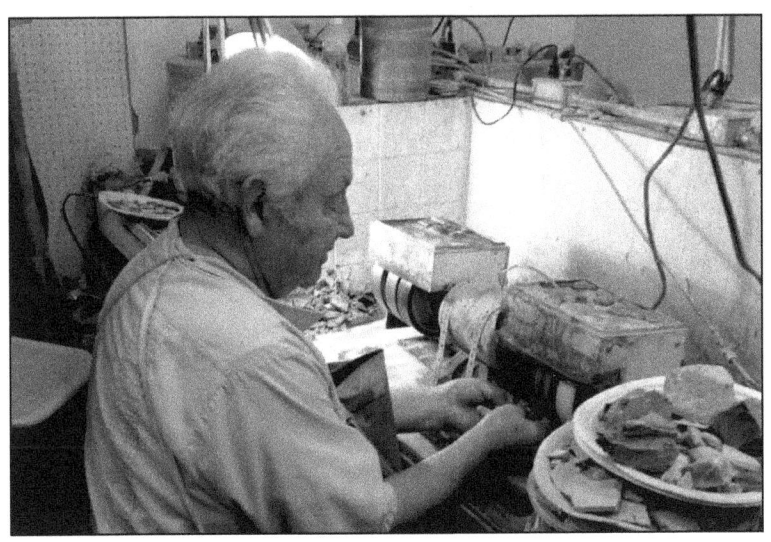

Robert at work in his Rock Shop
(Yelp Photo by Pat B.)

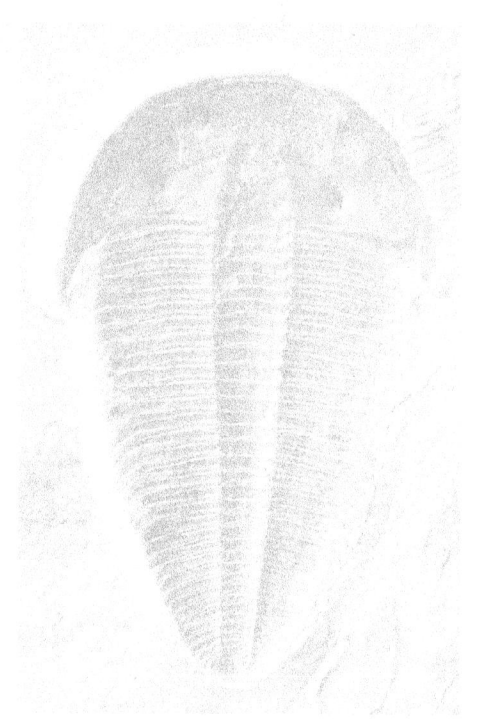

About the Author

Mark Harris and his wife Luree are the parents of five children. Both Mark and Luree taught school in California for many years. Mark's major teaching assignment was high school biology. He has a basic knowledge of most disciplines of science. While attending the University of Utah he took several classes in geology, his favorite professor being Richard A. Robison. Mark currently resides in Sandy, Utah.

Mark has collected family stories and pictures during most of his adult life and has written about various family members. Over the past fifteen years he has self-published more than 20 books, many relating to family members. He has also published calendars that display his photographs that cover a variety of fields. He is pleased to have this current book that highlights the trilobite collection of his brother Robert.

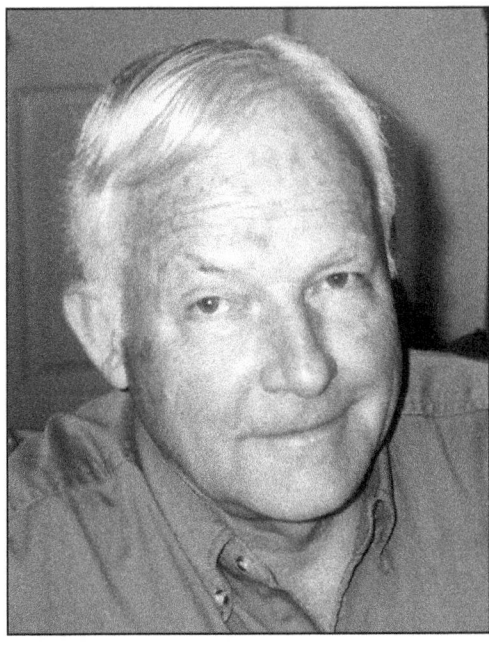

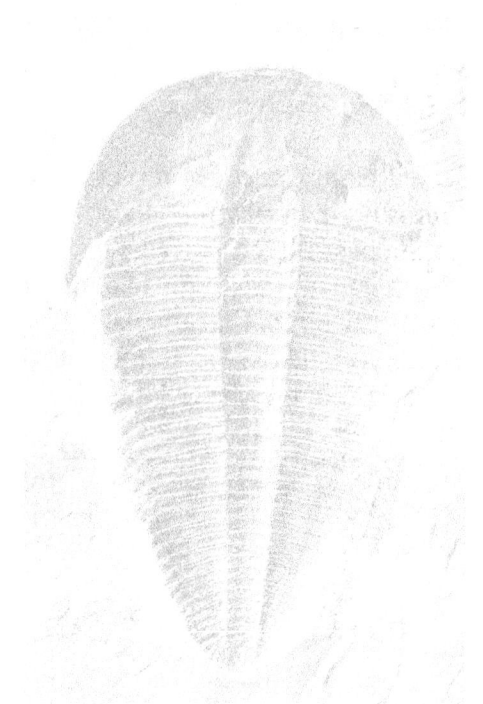

Research Sources

The following sources have been helpful in obtaining information and identifying the trilobites pictured in this publication.

Books:

Exceptional Cambrian Fossils from Utah
A Window into The Age of Trilobites
by Richard A Robison, Loren E Babcock and Val G Gunther

The Back to the Past Museum Guide to Trilobites
by Enrico Bonino and Carlo Kier

Websites:

westerntrilobites.com

fossilmuseum.net

fossilmall.com

Great Basin Museum website: Delta, Utah

UTAH TRILOBITES

Printed in the USA
CPSIA information can be obtained
at www.ICGtesting.com
LVHW012151130124
768930LV00010B/778